Control Things with Your Timex Sinclair

Control Things with Your Timex Sinclair

Robert L. Swarts

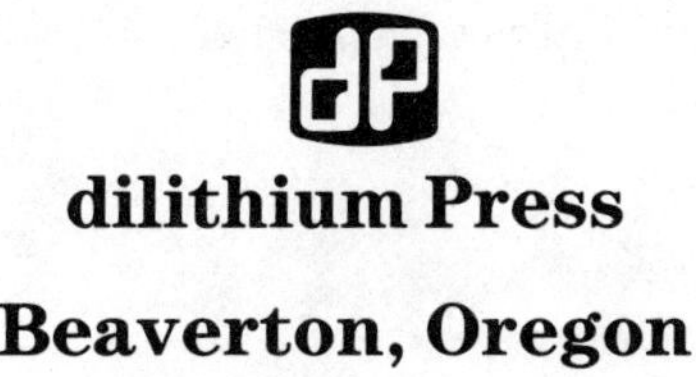

dilithium Press

Beaverton, Oregon

10 9 8 7 6 5 4 3 2 1

Library of Congress Cataloging in Publication Data

Swarts, Robert L., 1942–
  Control things with your Timex Sinclair.

  Includes index.
  1. Automatic control—Data processing. 2. Timex 1000 (Computer) I. Title.
TJ213.S84 1984   629.8′95   83-19016
ISBN 0-88056-127-0 (pbk.)

Cover: Eric Haag

Printed in the United States of America

dilithium Press
8285 S.W. Nimbus
Suite 151
Beaverton, Oregon 97005

Table of Contents

An Important Note

The publisher and the authors have made every effort to ensure that the computer programs and programming information in this publication are accurate and complete. However, this publication is prepared for general readership, and neither the publisher nor the authors have any knowledge about or ability to control any third party's use of the programs and programming information. There is no warranty or representation by either the publisher or the authors that the programs or programming information in this book will enable the reader or user to achieve any particular result. No liability is assumed for damages resulting from the use of the information herein.

Introduction

You say you're tired of games, you've plotted your biorythms until they've become arythmic, you know your time payments to ten decimal places, and the IRS can't possibly find anything wrong with your tax return this year? Now you're just screaming for something new and different but, after all, just how much more can you expect from a $50 "toy" computer?

Well, let's take that a step at a time. Is your Timex/Sinclair (TS) 1000 (ZX80, ZX81, or MicroAce) really a "toy"? Compared to a full-blown Apple II+ it may appear toylike, but it's several times faster and considerably more powerful than the 18,000 vacuum tube ENIAC of three decades ago—and that cost millions! Somewhat more recently you could have spent $400 for one of the original HP35 calculators. It, too, was slower than your TS 1000 and couldn't be programmed, but no one even thought about calling *that* a toy. Let's face it—you've got a real computing machine on your hands, albeit one with a few shortcomings. Whether you choose to consider it a toy depends more on what you do with it than on what it actually is. Fight off the alien invaders, and you've got a toy; control your home heating for maximum economy, and you have a very useful tool indeed.

How much usefulness can you expect? Quite a lot. There already exist numerous peripheral and add-on devices including extra memory, printers, mass storage devices, sound and voice synthesizers, light pens, data collection and control subsystems, high-resolution graphics packages, and telephone modems specifically tailored for the TS 1000. There are also hundreds of other add-ons that you can adapt to work with your computer if you have the necessary electronics know-how. Unfortunately, many of these products cost more than the computer itself and are somewhat less convenient to use than the ones designed for more expensive machines. If you were willing to spend the money and put up with a few extra key strokes, you could make the TS 1000 do most of the things that the bigger computers do. But then, you probably wouldn't have bought a $50 computer in the first place.

This book is written specifically for people like you who want to make their computers do useful things—to interact with the outside world in some sense—but not take out a second mortgage to do so. Sure, you'll have to do a little work, and the bits and pieces you'll assemble won't be *quite* as general in application as those you might purchase, but they'll cost only about $10 each instead of $100. And you may even learn a little in the process.

Some of the projects you'll consider are building blocks that can be used, either as is or in slightly modified form, to construct more sophisticated systems more applicable to your needs. Others represent minimum configurations; your own ingenuity can be used to embellish them. (For instance, you'll learn how to build a single-scale voltmeter and find some hints on how to make it multirange, but you'll have to complete the job yourself).

This book assumes that you understand simple BASIC programs and have very little experience with **PEEK**ing, **POKE**ing, or machine coding. The projects do not require vast technical knowledge, although the ability to solder, to read simple schematic diagrams, and to work with machine coding are all assets. (Incidently, if you plan on getting the full benefit of your computer, you'll have to learn machine coding somewhere along the line, and now is as good a time as any.) A number of excellent books on assembly and machine language programming are available at most computer shops, but Chapters 1–4 show you some of the essentials you'll need in order to understand the programs in this book. Of course, if you know the material already, you can safely skip Chapters 1–3, but you should read Chapter 4. It identifies the machine code commands used in this book, it contains a handy machine code loader, and it explains the notation used to list the machine code mnemonics.

Although all of the programs in this book function as stand-alone units, you'll use the subjects developed in the earlier chapters when you get to the later ones. If you wish to understand the programs rather than merely copy them, the best plan is to start at the beginning and work your way through the entire book.

You'll have to live with some limitations intrinsic to this computer. Clive Sinclair, the English gentleman responsible for all the ZX80-derived computers, succeeded in keeping costs minimal by taking some shortcuts. For example, the computer's memory addresses are not fully decoded. If you **PEEK** 33241 you are also **PEEK**ing 473. You can correct this only with external circuitry as the manufacturers of 56K or 64K memories do. Addresses between 8K and 16K are not available for BASIC programming and you can access them only by **PEEK**ing and **POKE**ing or through machine code. The display file cannot straddle the boundary between the lower 32K and the upper 32K of memory, or the system will crash. Elaborate programming steps must be taken to circumvent this problem if you wish to write a BASIC program longer than 16K. As long as you add 16K of memory or less, these limitations will not affect you. Others may. There are no **IN** or **OUT** commands in TS 1000 BASIC. You accomplish I/O (input/output) in machine code or by memory mapping the port and then **PEEK**ing and **POKE**ing it. Your keyboard is scanned rather than encoded, a slower method that eats up some of your potential I/O ports.

The character codes the Sinclair computers use are non-standard. Before your machine can communicate with most other devices, you will have to convert its codes, either in hardware or software, to ASCII (the American Standard Code for Information Interchange). Most importantly, unlike more expensive computers, which use a special circuit to control the video display, the TS 1000 forces the Z80 microprocessor itself to do the job. As a result, when you're in the SLOW mode the microprocessor spends about 5/6 of its time displaying and only 1/6 of its time computing. Your user's manual may say the microprocessor spends only 75%–80% of its time displaying—and this may be true for UK machines. (Sinclair designed its computer according to the United Kingdom's TV standards, which are different from ours.) By actual measurement, it takes six times longer to run any routine in SLOW mode than in FAST on the American versions of the computer. Try a short program as proof, timing the execution in both modes with a stopwatch or the second hand of a watch or clock. You should never expect to do anything very rapidly in SLOW mode, and you should always expect it to make timekeeping operations more difficult. Despite these limitations, you'll still find the TS 1000 a very versatile and useful machine capable of many more tasks than the owner's manual might lead you to think.

You can buy all of the parts you'll need for the projects in this book at Radio Shack stores. This doesn't mean that Radio Shack is the best, the only, or even the cheapest supplier. But the chain *is* ubiquitious, and knowing where to obtain parts easily will save you time.

The author assembled and tested all of the projects, hardware and software, with a ZX-81 or TS 1000, some with a MicroAce equipped with 8K ROM (*Read-Only-Memory*). None of the projects require more than 2K of memory, and almost all will fit in 1K. However, if you plan to use your computer in the indefinite future and wish to enjoy it as much as possible, then expanding your computer's memory by at least 16K should be a fairly high priority. If you're contemplating handling large amounts of data, large arrays, or possibly some machine-coded utilities, then you should think seriously about a 32K–64K expansion.

Chapter 1

Unbaffling Bits

Certain loosely connected subjects like bytes, memory structure, binary math, and bits are often surrounded by mystery for the novice. In order to eliminate some of the confusion, this book includes some tutorial chapters. This particular chapter sets out to unbaffle as many bits as possible.

It will not treat its subjects in great depth. Instead, it will give you just enough knowledge to understand the programs that will develop in the remainder of the book. Your local library or computer store can supply you with numerous texts on binary math, flow charting, and machine coding. You can find useful information on TS 1000 memory structure, the display file, storing machine code, and the **PEEK**, **USR**, and **POKE** functions/commands in your owner's manual.

Let's begin by considering the concept of ''number.'' Number generally denotes a quantity of something. It doesn't matter what that something is; it could be apples, oranges, the change in your pocket, or the miles to the nearest star. Nor does number have a unique name. In fact, the *concept* of number really does not require a name. Suppose you saw a man carrying a small sack and and wanted to know what was in it, but could not speak his language. You might point to the sack and raise your eyebrows in question. The man might pick up a small stone, hold it out to you, and point to the sack. You might not be sure, but you'd probably guess that his sack had stones in it. You could then hold out one finger, then two, then three, shrug and raise your eyebrows, signifying you wanted to know how many stones. He might smile and hold out both hands with fingers spread. He might even tick off three or four fingers and then spread both hands. Even if you weren't sure what his sack contained, at this point you'd be certain he had ten of whatever it was. Had he had no fingers, he might have made ten marks in the dirt. The important point is that he would have conveyed a concept of quantity, not particular things, not names of quantities.

Similarly, if you spoke German and English, you'd know that "thirty" and "dreizig" both mean "30" even though they have only one letter in common. It's the same with the shorthand we use to express numbers. A "17" in decimal (our everyday counting system) and a "10001" in binary both denote the quantity seventeen.

In common speech, we frequently refer to the notation we use to represent a quantity as a "number," and as you continue reading this book you should understand the word "numbers" to mean the string of digits or numerals you would normally write for a given quantity. What does this "notation" mean? Take a look at this example.

$$
\begin{array}{rcll}
1352 & = & 1*1000 & = \quad 1*10**3 \\
      &   & +3*100 & \quad\quad +3*10**2 \\
      &   & +5*10 & \quad\quad +5*10**1 \\
      &   & +2*1   & \quad\quad +2*10**0 \\
      &   & \overline{} & \quad\quad \overline{} \\
      &   & 1352   & \quad\quad 1352
\end{array}
$$

Remember that any number raised to the zero power is one. Now try the number 104.

$$
\begin{array}{rcll}
104 & = & 1*100 & = \quad 1*10**2 \\
    &   & +0*10 & \quad\quad +0*10**1 \\
    &   & +4*1 & \quad\quad +4*10**0 \\
    &   & \overline{} & \quad\quad \overline{} \\
    &   & 104   & \quad\quad 104
\end{array}
$$

So how does this work in binary? Much the same way.

$$
\begin{array}{rcll}
10001 & = & 1*16 & = \quad 1*2**4 \\
      &   & 0*8 & \quad\quad +0*2**3 \\
      &   & 0*4 & \quad\quad +0*2**2 \\
      &   & 0*2 & \quad\quad +0*2**1 \\
      &   & +1*1 & \quad\quad +1*2**0 \\
      &   & \overline{} & \quad\quad \overline{} \\
      &   & 17   & \quad\quad 17
\end{array}
$$

The root words of decimal and binary mean ten and two respectively, and these numbers are the "bases" of the decimal and binary numbering systems. Our numerical shorthand makes each digit in our strings the coefficient of the base to an integer exponent. Call the base of the number system b and, starting from the right, call the digits in the string d0, d1, d2, . . . dN. Then, for any number system

$$
\begin{aligned}
\text{dN}\ldots\text{d2 d1 d0} \quad = \quad & \text{dN}*\text{b}**\text{N} \\
& \vdots \\
& +\text{d2}*\text{b}**2 \\
& +\text{d1}*\text{b}**1 \\
& \underline{+\text{d0}*\text{b}**0} \\
\end{aligned}
$$

decimal equivalent

Converting from decimal to binary is just about as easy. Start by finding the largest power of two that will divide the decimal number once but not more than once. Then divide the rest of the remainders by successively smaller powers of two as in the example that follows.

In converting 75 decimal to binary, $64(2**6)$ is the largest power of two that will divide 75.

75/64 = 1, remainder 11	binary = 1
11/32 = 0, remainder 11	binary = 10
11/16 = 0, remainder 11	binary = 100
11/8 = 1, remainder 3	binary = 1001
3/4 = 0, remainder 3	binary = 10010
3/2 = 1, remainder 1	binary = 100101
1/1 = 1, remainder 0	binary = 1001011

The binary equivalent of a decimal number is the string of binary digits that results from the repeated division.

Counting in any number system is essentially the process of repeatedly adding one to a number.

$$1+1=2, \ 2+1=3, \ 3+1=4$$

This works fine until you get to the number that's equal to the base. Here are two familiar examples in base ten.

<table>
<tr><td align="right">9
+ <u>1</u>
0</td><td></td><td align="right">999
+ <u>1</u>
0</td></tr>
<tr><td>carry <u>1</u>
10</td><td></td><td>carry <u>1</u>
00</td></tr>
</table>

$$\begin{array}{r} \text{carry } 1 \\ \hline 000 \\ \text{carry } 1 \\ \hline 1000 \end{array}$$

The very same rules apply in binary. Remember, two is the base.

$$\begin{array}{r} 1 \\ +\ 1 \\ \hline 0 \\ \text{carry } 1 \\ \hline 10 = 2 \end{array} \qquad\qquad \begin{array}{r} 111 \\ +\ \ \ 1 \\ \hline 0 \\ \text{carry } \ \ 1 \\ \hline 00 \\ \text{carry } 1 \\ \hline 000 \\ \text{carry } 1 \\ \hline 1000 = 8 \end{array}$$

As you can see, a binary digit, or bit (from *binary* and dig*it*), can have only the value one or zero, whereas a decimal digit can range from zero to nine. Note that all of these examples work just like the odometer in your car. The units wheel (for single miles) turns 10 times for each increase in the tens wheel (for ten-mile increments), and the tens wheel turns ten times for each increase in the hundreds wheel (for hundred-mile increments). If you had a binary odometer, the units wheel would turn only twice before the twos wheel turned once, and the twos wheel would turn only twice before the fours wheel turned.

Imagine you have an eight-digit binary odometer on a brand-new car. Its mileage will increase just as the binary numbers increase in Table 1.1. Notice that the leading zeros mean nothing. Whenever a single one shows up in some position, the pattern of digits to its right repeats in successive numbers the pattern of all the earlier numbers. Whenever a single one shows up, the decimal value is a single-integer power of two. The largest number you can represent with 8 bits is 255; after that the odometer clicks over and repeats the original pattern. The odometer did carry a one, but had no place to put it. To get a feel for binary addition, try generating the first 32 binary numbers by starting at zero and successively adding one to the previous number:

TABLE 1.1 Eight-Digit Binary Numbers and Their Decimal Equivalent

decimal	binary	Powers of two
0	00000000	—
1	00000001	2**0
2	00000010	2**1
3	00000011	—
4	00000100	2**2
5	00000101	—
6	00000110	—
7	00000111	—
8	00001000	2**3
.	.	.
.	.	.
.	.	.
15	00001111	—
16	00010000	2**4
.	.	.
.	.	.
.	.	.
127	01111111	—
128	10000000	2**7
.	.	.
.	.	.
.	.	.
254	11111110	—
255	11111111	—
256	00000000*	2**8

*Because the binary numbers column of this table can contain only eight digits, 100000000, the binary equivalent of 256 decimal, appears as all zeros. The one carries over, but does not show up in the binary column.

$$\begin{array}{r} 0 \\ +\ 1 \\ \hline 1 \\ +\ 1 \\ \hline 10 \\ +\ 1 \end{array}$$

and so on. If you can do that, then you should be able to add any two binary numbers.

$$\begin{array}{r} 01100101 \\ +01001100 \\ \hline 10110001 \end{array} \qquad \begin{array}{r} 00101011 \\ 01011001 \\ \hline 10000100 \end{array}$$

Try several more examples on your own. You can convert them to decimal to check your addition.

So now, how do you subtract in binary? You could do it just as you do in decimal, by borrowing one from the digit to the left when necessary, but this is not how home computers do it. Instead, like a computer, you add the negative of the number you want to subtract to the number you want to subtract it from. This is a familiar operation in decimal. There's no difference between subtracting the positive and adding the negative.

$$\begin{array}{r} 19 \\ -(+6) \\ \hline +\ 13 \end{array} \qquad \begin{array}{r} 19 \\ +(-6) \\ \hline +\ 13 \end{array}$$

However, there is a small problem in deciding how to represent a negative binary number. Assume that any time the first digit on the left is a one you'll call that number negative. From the list of binary numbers in Table 1.2 you'll see that you could then represent numbers from −127 to +127 with eight digits. But does this give the right answer? Alas, no.

$$\begin{array}{r} 16\ = \\ +(-8)\ \\ \hline +\ 8\ \end{array} \begin{array}{l} 00010000 \\ 10001000 \\ \hline 10011000 = -24 \end{array}$$

What you'll need to do is use something called the two's complement of the number. To do the projects in this book, you won't need to understand why this works; you'll just need to know how to do it. Many books on computers and all books on binary math describe the two's complement in some detail, and you can consult them if you want more information.

To begin, find the complement (the one's complement) of a number by changing each one to a zero and each zero to a one.

number	(one's) complement
10001101	01110010
00101010	11010101
11111111	00000000
10101010	01010101

To get the two's complement, add one to the complement.

number	(one's) complement		two's complement
10001101	01110010	+1	01110011
00101010	11010101	+1	11010110
11111111	00000000	+1	00000001
10101010	01010101	+1	01010110

Now repeat the subtraction problem.

$$
\begin{array}{rl}
16 & 00010000 \\
+(-8) & 11111000 \\
\hline
+8 \quad 1 & 00001000 = +8
\end{array}
$$

The carried over one is ignored. Here's another example.

$$
\begin{array}{rl}
72 & 01001000 \\
+(-100) & 10011100 \\
\hline
-28 & 11100100 = -28
\end{array}
$$

Notice that the result of the binary addition is in two's complement notation, so you must convert it back to make sure that it is, in fact, -28.

$$\text{complement } 11100100 = 00011011 \quad +1 = 00011100 = 28$$

You know it *is* -28 because the first digit on the left was a one. Here, then, is the rule for subtracting B from A.

$$
\begin{array}{ccc}
\begin{array}{r} A \\ -B \\ \hline C \end{array} & = & \begin{array}{r} A \\ +(\text{two's complement B}) \\ \hline \text{(ignore carry)} \qquad C \end{array}
\end{array}
$$

TABLE 1.2 Eight-Digit Binary Numbers Represented in Two's Complement Notation

| Positive | | Negative | |
Decimal	Binary	Decimal	Binary
127	01111111	−128	10000000
126	01111110	−127	10000001
125	01111101	−126	10000010
⋮	⋮	⋮	⋮
16	00010000	− 16	11110000
15	00001111	− 15	11110001
.	.	.	.
.	.	.	.
.	.	.	.
8	00001000	− 8	11111000
7	00000111	− 7	11111001
6	00000110	− 6	11111010
5	00000101	− 5	11111011
4	00000100	− 4	11111100
3	00000011	− 3	11111101
2	00000010	− 2	11111110
1	00000001	− 1	11111111
0	00000000		

Table 1.2 shows the numbers you can represent with eight bits if you use two's complement notation. Note that you can still work with 256 numbers, but that they now range from −128 to +127.

Here are a couple more points worth remembering. Since the rule for finding the two's complement of a number, N, was

$$\text{one's complement} + 1 = \text{two's complement}$$
$$\text{of N} \qquad\qquad \text{of N}$$

as Table 1.2 shows, you could just as well have written

$$\text{one's complement} = \text{two's complement}$$
$$\text{of N} - 1 \qquad\qquad \text{of N}$$

Because −1 is represented as all ones, it will generate a carry when it is added to any number except zero. You'll make use of this characteristic later. Now, try adding +1 to +127. Do you get −128?

How about −1 to −128 for an answer of +127? Do you see what is happening? The upper and lower entries of the table follow one another just as though they were written on a cylinder (like an odometer). This really shouldn't surprise you since the same thing happened with positive numbers when we added 1 to 255 to get zero. If you're working with 8-bit numbers, it's worth remembering that whenever the solution to a problem is greater than +127 or less than −128, your computer may return an erroneous answer.

Does this mean you can deal only with numbers in this range? No. All you have to do is keep adding spaces to the left of the existing numbers. A place, then, is available for the carry or for larger numbers. If you add another 8 bits, the biggest number you can handle becomes

$$2**15+2**14+\ldots 2**1+2**0 = 65535 = (2**16)-1$$

The largest number you can count becomes

$$(2**\text{number of digits})-1$$

Or, if you choose to use two's complement notation, the largest number becomes

$$(2**(\text{number of digits}-1))-1$$

and the smallest becomes

$$-(2**(\text{number of digits}-1))$$

While adding 8 bits doesn't solve the problem of erroneous answers it vastly increases the size of the numbers you can handle. Actually, no solution will yield an exact answer when the numbers are arbitrarily large if you restrict yourself to a finite number of digits. This is true even in decimal—try it. Your computer—most computers—will handle much larger numbers, but the numbers will no longer be exact. To handle larger numbers you have to abandon integer arithmetic (the system we have been working with) and use floating point notation. This scheme keeps track of only a predetermined number of the most significant digits, those on the left of the expanded notation.

Although it's not exactly the way the TS 1000 carries it out, the following example will illustrate how floating point notation works. Suppose you have to keep track of the number 113,456,789 but you really don't care whether it's rounded to 113 million or 114 million. Here's how you do it with 16 bits. Let the first 8 bits represent any number between 0 and 255. Then let the second 8 bits represent the power of ten necessary to make the number the correct magnitude.

$$113,456,789 = \text{approximately } 113*10**6$$

You can represent the latter term as

Mantissa	Exponent
113	6
01110001	00000110

As you can see, you could represent numbers up to $255*10**255$ (255 followed by 255 zeros) by using this technique, but your numbers could never be more accurate than one part in 255. Your Timex Sinclair computer actually uses 32 bits for the mantissa and 8 bits for the exponent.

Division and multiplication are not hardware functions of the Z80 microprocessor; they are carried out by machine code programs stored in your computer's ROM (*Read-Only-Memory*). Since you don't need to use these operations in any of the projects in this book, all you have to know is that your computer does them in very much the same way you do them on paper. The only significant difference is that the bits can have only the value 0 or 1. Since one times anything is still the same anything and zero times anything is 0, the operations can be carried out by repeated shifts of position and additions of numbers. Try multiplying binary 8 by binary 7.

```
    1000 =  8
     111 =  7
    ─────    ─
    1000
   1000
  1000
  ──────
  111000 = 56
```

To find the total you can simply add three shifted binary eight's.

But why do computers use binary math, and why has this chapter concentrated on 8- and 16-bit numbers? It's easy to build electronic circuits that can only be either on or off. Frequently these two states are referred to as high and low or as true and false, but since there are only two states they can be just as easily represented by a one and a zero. It is also relatively easy to arrange such digital circuits as counters and logic blocks, and these blocks can carry out binary math. An "AND" circuit, for instance, requires that both of its inputs be high (call high = 1) for its output to be high while an "Exclusive OR" circuit requires that one and only one of its inputs be high for a high output. These circuits could be used to add two single digit binary numbers. Suppose that you call the binary numbers A and B, and suppose that each is represented by a wire that may be at a high voltage or a low voltage depending upon whether the number is high or low. Both these wires go to both inputs to both circuits. In Table 1.3, which is often called a truth table, the outputs of the circuits as a function of the states of A and B appear. Notice that the outputs of the circuits correctly show the sums of the two bits. Of course, your computer is far more complex, but you get the idea.

This discussion has concentrated on 8- and 16-bit numbers because that is what most home computers—including your TS 1000—use. When working with binary numbers, it's convenient to have the number of bits equal to a power of two, and it's advantageous to use as large a number as possible. The microprocessor treats the groups of bits as discrete units, regarding them as either instructions or data. It will regard them as instructions unless a prior instruction has told it to consider them data (numbers). Many more powerful scientific and industrial computers deal with larger groups of bits, 32 or even 64 at a time. The next chapter discusses how a computer manages these groups.

TABLE 1.3 Truth Table for AND and Exclusive OR Circuits

A	B	AND	Exclusive OR
0	0	0	0
1	0	0	1
0	1	0	1
1	1	1	0

Chapter 2

Know Thy Computer's Memory

Imagine, if you will, a rather strange office fitted with cabinets that have slots for 65535 small file drawers. Each slot is numbered in order, the numbers starting at 0 and progressing to 65535. Some of the slots have file drawers in them, and some do not. Further, some of the installed drawers have glass tops so that you can read what they contain but can't deposit anything, and still others have signs that say ''reserved for office operations information.'' This is very much what the TS 1000 memory looks like.

Before you can see how this filing system works, you need to look at the Z80 microprocessor (mP) that's the central office—the heart of the TS 1000. Z80s and many other microprocessors possess a data bus, an address bus, and a control bus. Physically these buses are sets of pins on the integrated circuit. They carry signals that you can categorize as high or low, true or false, or 1 or 0. The mP treats each set of signals on the data and address buses as individual units. The control bus signals are somewhat more autonomous. Since the signals on the data and address buses can only be on or off, the sets of signals may be in the form of binary numbers.

If you were a secretary in our hypothetical office, the office manager might say something like ''Go to 13456 (using the address bus), fetch (using the control bus) the information in it, and bring it to me (via the data bus).'' He would, however, tell you in code. The code could be 213,13456 accompanied by a bell or a buzzer, all of which you would interpret as

213 transfer information to/from the address
following this code
bell transfer to the file
buzzer transfer from the file
13456 address following the code

Now there's another peculiarity about the office file drawers; the information in them is in binary code, and the length of the code is limited. The word size of the system defines the length of the code stored in each drawer, and each drawer contains one and only one word of data. Obviously, if you wish to convey much information, you're going to have to use more than one drawer. Because you have to go to different drawers and retrieve the information sequentially, it will also take longer to assemble the message. Usually, the width of the data bus, in number of bits, will be equal to the word size of the memory. So, as you can see, the primary advantages of using larger word sizes are that more information can be contained in each word, and that retrieval of that information will be faster since you need not fetch data from memory as many times. The disadvantage is that larger file drawers are expensive to build. You need to reach a compromise wherein the files are big enough to do the job in a reasonable time and yet cheap enough to build so that you can afford them. In all current home computers the word size is either 8 or 16 bits. In the TS 1000 it is 8 bits.

It sometimes happens, especially in 16-bit and larger machines, that you want to look at just part of a word. This is true, for instance, when you want to examine the codes for your keyboard keys. Since all of the keys are encoded in 8 bits, a 16-bit word can hold two codes. In a 16-bit system you frequently treat the first and last 8 bits of the word as two separate units called bytes. Through common usage byte has evolved to mean any group of bits treated as a unit by the mP. In most modern mP-based 8-bit computers, including the TS 1000, the word and byte size are both 8 bits.

When you do break a piece of information—say a 16-bit number—into 2 bytes, you categorize the first and last 8 bits as high-order and low-order bytes. The high and low refer to the powers of two represented by the byte, the left-hand byte corresponding to the higher powers, and the right-hand byte to the lower powers. Thus, you can speak of the high- or low-order byte of the address bus, which, in the case of the Z80, is 16 bits long. Note, though, that the Z80 itself does not split the address bus, but treats it always as one 16-bit number. Not all mPs handle their address buses this way.

Now that you understand that you have an office-like system for storing and retrieving information, you'd probably like to get information into and out of it. One way to do so would be to put a printer in the hallway outside the office and run its cable to one of the empty file drawer slots. Then, whenever you deposited something in that slot, it would go to the printer rather than stay stored in the slot. Similarly, the printer you arrange to put in the hallway could have a keyboard whose output is fed to a device that stores the code corresponding to the last key you press. This device would also occupy one of the unused slots. Then, whenever you retrieved information from that slot, you'd be getting the key code from the keyboard. Inputting or outputting information in this fashion is termed memory mapped I/O (*Input/Output*) because the input or output device or its controller circuit is treated as though it were a memory location. Many of the peripherals you can purchase for the TS 1000 use this technique, but this book will show you how to use another.

The Z80 is unique as 8-bit mPs go in that it has a special set of input/output commands that can be used to directly control up to 256 I/O devices. Imagine that your office has 256 additional slots numbered 0 to 255, slots into which you can plug I/O devices. These slots are separate from the first 256 slots in your filing system. When the office manager flashed a light (a control bus function), the secretary would go to one of these addresses rather than to one of the first 256 addresses of of the filing system to deposit or receive information. The advantages of this technique are that it does not consume any file space, it is slightly faster in execution, and it takes less space to store the command (the address is encoded in 8 rather than 16 bits). In addition, the parts (address decoders) of the I/O devices that determine whether the manager is calling their address or some other address can be less complex (again because the system uses 8-bit addressing). The disadvantage of this setup is that you can't issue I/O commands from BASIC in the TS 1000. Figure 2.1 is a block diagram of a typical Z80-based computer showing the memory, I/O devices, and buses. (You'll find an explanation of the clock function later in this book.)

Inside your TS 1000 are two integrated circuit (IC) memories. The first is a Read Only Memory (ROM), which is analagous to the

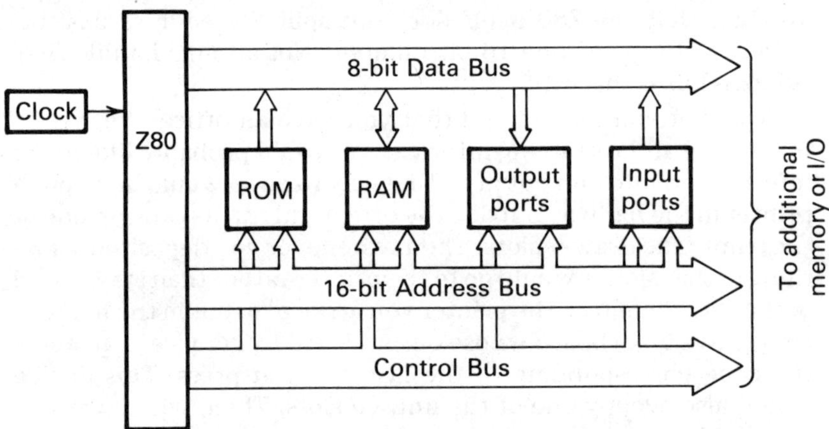

FIGURE 2.1 Typical Z80-Based Computer

group of file drawers with the glass over them. The mP can read
the codes contained in this group but altering them would destroy
the device. The particular ROM in the TS 1000 occupies addresses
0–8191; that is, it contains the file drawers with addresses
0–8191. In computer jargon, these locations are the first 8K of
memory. The K signifies 1000 and 8 is just an approximation of
8.191. In the same way, you might speak of 16K instead of 16383,
or 64K instead of 65535.

Why do computers have memories that can only be read? Because those same memories can also permanently retain whatever is stored in them even when the power is off. In most home
computers what is stored in these memories is the set of instructions that make the computer behave as a BASIC-speaking unit
and the "utilities," or subroutines, needed to control such fundamental functions as displaying data, loading and saving tapes,
interpreting the keyboard, and so forth.

The second memory IC is a Random Access Memory (RAM). It
contains general-purpose file drawers. You can take information
from these drawers or store information inside them. The particular 2K memory chip installed in the TS 1000 occupies locations
16384–18431.

Your TS 1000, as it comes out of the box, has the memory
locations remaining above RAM from 18432 to 65535 and the
block between ROM and RAM, 8K to 16K, all vacant (there are no
file drawers). Figure 2.2 shows a diagram of the total potential
memory, called a memory map. You can use the block of memory

between 18431 and 65535 and the block between 8K and 16K for more ROM, RAM, or memory mapped I/O, but there is a problem. Because Mr. Sinclair did not originally intend that anyone add memory anywhere except in the 16K–32K area, the addresses of memory locations within the computer are not fully decoded. As a result, the decoding circuits examine not the whole address word but only the bits needed to activate the device the decoder is attached to. As an example, tell your computer to **PRINT PEEK 7000** and then **PRINT PEEK 15192**. The same value, right? Now notice that the difference between 15192 and 7000 is 8192, which is $2**13$, or bit 13 of the address bus. The decoding circuit for ROM doesn't look at bit 13; it's been told that as long as bit 14 is —

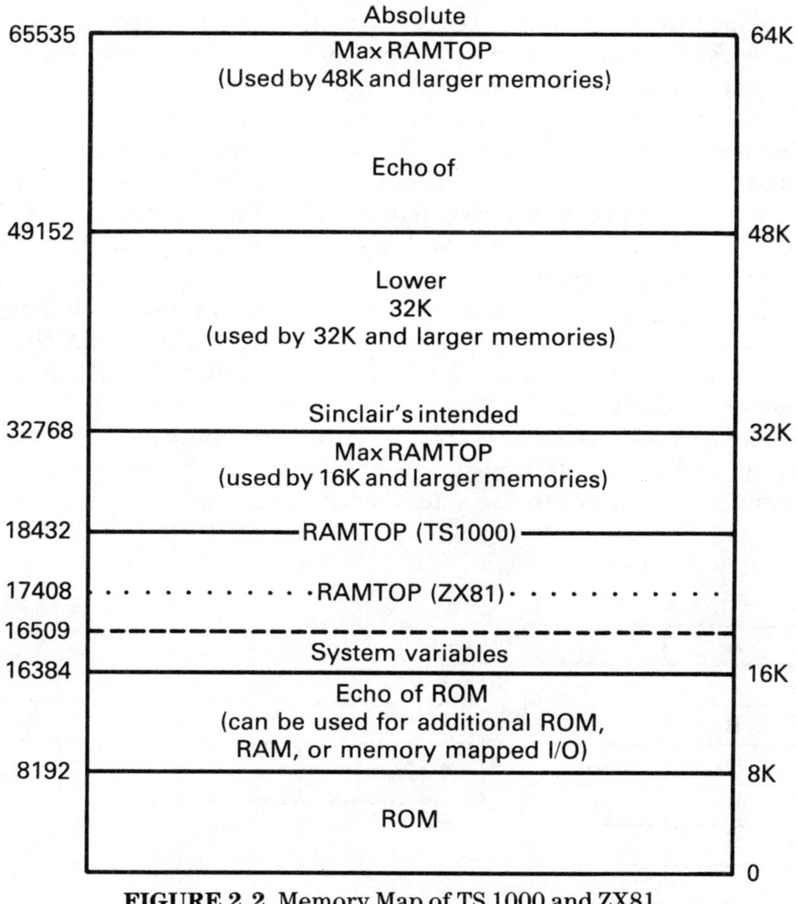

FIGURE 2.2 Memory Map of TS 1000 and ZX81

not set, the mP is trying to access ROM. Therefore, locations 8192–16383 "echo" every location in ROM. The same thing happens above 32K for similar reasons. The upper 32K of memory echoes all the lower 32K. In fact, it is the echo of RAM that drives the video display. (Bit 15 of the address bus is set during a display.) Fortunately for the manufacturers of add-on equipment—and for users like you—it's possible to rectify the decoding problem externally and make the whole memory map available for use.

Before getting any further into memory, let's take a look at the inside of the mP. The Z80 contains 22 internal registers. You'll need to consider only the nine that you'll use later in the book. (See Figure 2.3.) These registers look and act for all the world like the file drawers in memory except that the mP can directly access them more rapidly and manipulate their contents in more ways than it can the external memory registers.

One of the nine registers, the program counter, or PC register, is 16 bits long. It contains the address of the memory location holding the code of the next instruction for the mP. It points into memory showing the mP where to go next. When power is turned on, it contains address zero. Normally, this register's contents will increment up through memory (0, 1, 2, . . .) unless the code contained in a particular memory location tells the PC to alter its contents. Such codes are called branch instructions, and they perform functions similar to **GOTO** or **IF . . . THEN GOTO** commands in BASIC.

The first 8-bit register is the A register, traditionally called the accumulator. It is the most powerful of all the registers; other registers are unable to use a number of the commands associated with it. For example, logical operations (COMPARE, AND, OR) must be carried out in A.

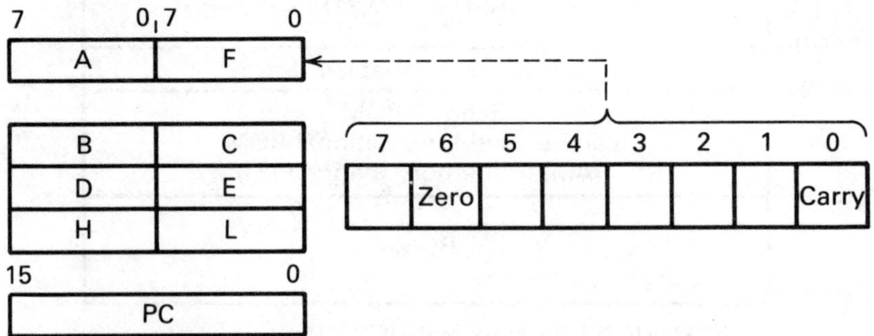

FIGURE 2.3 Nine Internal Registers of the Z80 Microprocessor

The F, or flags, register is 8 bits long but the mP doesn't really treat it as a single unit. Rather, the outcome of certain operations determines how the individual bits in it are set or reset. If an addition results in a zero sum, the zero bit is set, and if a carry is generated, then the carry bit is set. In both cases the mP waves a flag. The zero and carry bits are the only two bits you need be concerned with.

The B, C, D, E, H, and L registers are general-purpose units. You can pair them to form 16-bit registers called BC, DE, and HL to facilitate 16-bit arithmetic or addressing. Some of the instructions associated with one or more of them are very powerful.

Consider now how you could use the whole system, first to add a number stored in 16800 to another number stored in 17001, and then store the sum in a location you input from a keyboard you've attached to I/O location 200. The instructions to do all this exist in Z80 machine code. Here's how they read in everyday language.

Fetch the first number from 16800 and put it in register B.

Fetch the second number from 17001 and put it in the accumulator, A.

Input the high-order byte of the storage location from I/O location 200 and put it in H.

Input the low-order byte of the storage location from I/O location 200 and put it in L.

Add A and B. (Keep the result in A.)

Store whatever is in A at the location the contents of HL give you.

As you'll see in Chapter 4, machine language programs consist of a string of codes corresponding to statements like these.

For now, though, let's return to memory. Because your computer has the built-in capability for different amounts of memory, one of the first things your TS 1000 will do is determine how much memory is actually installed. When you turn on the power, a hardware-generated reset pulse will force the contents of the PC to zero. The instructions beginning at zero turn off the SLOW display and cause the mP to load the value 2 into all memory locations from 16384 to 32767. Naturally, if the memory isn't there, the number isn't stored. The computer checks its memory locations sequentially until it finds the first memory location that doesn't contain 2 (the first nonexistent memory cell, or empty file drawer slot). It then deposits the location's address in system-variable RAMTOP. The computer now knows how much memory it has.

If you add a 16K memory pack, then locations 16384–32767 will contain functioning memory cells, and RAMTOP will be set at 32768. If you add on more than 16K of memory, then you will have to **POKE** in the value for RAMTOP since the TS 1000 doesn't write twos into memory above 32767 or check the block of memory between 8K and 16K. Note that a memory device with a bad storage location within the 16K–32K block will result in a RAMTOP value less than the amount which should be obtained for the installed memory. This can be a valuable trouble-shooting hint if your memory doesn't behave as you think it should.

The programs you write are stored in RAM, but not all of the installed RAM is available to your for program storage. Remember those file drawers marked ''reserved to office operations information''? In a real office they might contain data on who is working at what task, when fees are due to be received, when bills must be paid and how much they are, when due dates for shipping occur, and a host of other information that an office manager needs if the office is to function smoothly. You can change some of the data (who is working on what task) at the cost of disrupting office operations, but altering some of the other data (when bills are due and how much they are) will mean running seriously afoul of the law and cause the business to ''crash.'' Locations 16384–16508, the system variables in Figure 2.4, play a similar role in the TS 1000. The locations contain parameters that tell the instructions in ROM what they are doing and how to find their way around in the useable memory above 16508. Although you can't store information in these 125 locations, you can read,

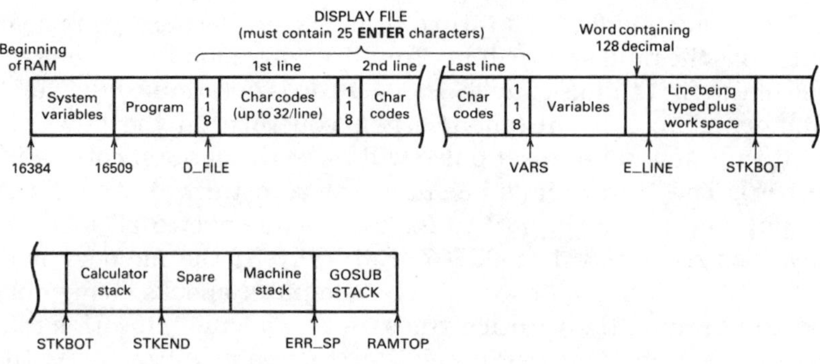

FIGURE 2.4 Memory Map of Installed RAM

and in some cases, alter the values of their parameters to design more efficient programs (restructuring your office for a particular task).

Other areas reserved to the computer are display file, variables, line being typed plus work space, calculator stack, machine stack, and **GOSUB** stack, but these areas vary in size according to the work the computer is doing and the nature of any program it is running. In principle, it's possible to determine the exact size of these areas at any time, but in practice it's far too complex a procedure. The best you can do is to make general observations. The fewer variables in a program, the smaller the variable space. The fewer nested **GOSUB**s, the smaller the **GOSUB** stack; the shorter the math algorithms, the smaller the calculator stack. Minimizing these items sometimes takes extra program lines, which themselves require more space. You will find the trade-off involved anything but clear-cut.

One of the most important attributes of these files is their flexibility. Not only do their lengths vary, but their starting locations are movable. Every time you add data to an area, all of the areas above that one must be pushed upwards in memory to make room for the addition. The one exception to this is the program area: it always starts at location 16509. The first item of the first line of a program, which will be the first digit of the line number, will always go here. You'll make vigorous use of this anomaly when you start writing machine code programs. Meanwhile, imagine that there's a program in the computer that consists of nothing more than two empty REM (remark) statements.

    10 REM
    20 REM

By consulting the owner's manual, you can determine where D__FILE (the beginning of the display file) is. Figure 2.5 is a slight modification of one in the manual. It shows that each and every BASIC program line starts with 2 bytes for the line number (even if the number could have been represented by 1 byte) and ends with one byte for the **ENTER** character, code 118. The 2 bytes following the line number are always present and contain the length of the text plus one for the **ENTER** character. The length of the text is the number of key strokes needed to type it plus 6 extra bytes for each number assigned in the text. Thus, every BASIC line carries with it a burden, or overhead, of 5 bytes of memory beyond the amount required to simply write the state-

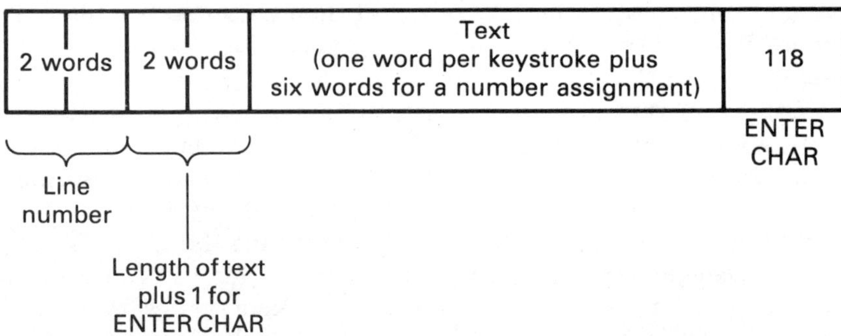

FIGURE 2.5 Format of a Program Line in TS 1000 Memory

ment. (That's why when you're using a computer with a small amount of RAM, you should try to squeeze as many operations as possible into each line.)

Once you know that the first byte of the program is in location 16509, you can count up six locations for each REM statement to discover that D__FILE will be in location 16521. If you add another line, the one-line program

    30 REM

you have to count up another six locations, which places D__FILE in 16527. All of the sections above D__FILE will be similarly pushed upwards. The reverse, of course, happens when a program line is deleted.

The display file is another area that you can often manipulate to advantage. It always begins with an **ENTER** character and contains a list of the codes for the characters being displayed on the screen. Another part of the ROM converts the codes into the patterns you see on your TV screen. The display can contain up to 24 lines of 32 characters. Each and every line ends with an **ENTER** character so that a full line in memory is 33 codes long. The shortest a line can be is one code, that of the **ENTER** character, so even if the display file is empty (nothing has been printed), it is 25 characters long (one for the first ENTER that begins the file and one for each of the 24 empty lines). A full file would contain $24 \times 33 + 1$ codes representing 24×32 characters. Even a space is a character, by the way, so if you print a space, the display file is not really empty even though the screen is blank.

You do not have control of the last two lines; the computer reserves them for the line you're currently typing or for inputting data. The display file is filled out from the left-hand screen edge, line by line. If you have a program that contains a simple print statement, the codes to be printed will immediately follow the beginning **ENTER** code. **PRINT** ''ABCD'' will result in a display file like this

ENTER,code A,code B,code C,code D,24 ENTERs

PRINT AT statements, however, jump to the line you want to print in, and they pad out the beginning of the line with spaces until the print position is reached. **PRINT AT** 10,12;''AB'' will yield a display file like this

10 ENTERS,11 space codes,code A,code B, 15 ENTERs

The computer skips the beginning **ENTER** and nine **ENTER**s representing the first nine empty lines, and arrives at the beginning of the tenth line. It then prints 11 spaces to arrive at the twelfth column where it prints the A and then the B. The rest of the display remains empty.

It is sometimes useful to fill up the portion of the screen you are going to use before you actually print in it so that you know the size of the display file and the place where the printing area starts. You can do this by using **PRINT** or **PRINT AT** commands containing only spaces.

```
10 PRINT ''          ''
20 PRINT ''          ''
```

fills up the first 10 columns of lines 0 and 1, and bytes 2–11 and 13–22 of the display file contain space codes.

```
10 PRINT AT 10,15;''   ''
20 PRINT AT 11,15;''   ''
```

fills up columns 0–15 of lines 10 and 11, and bytes 11–25 and 27–41 contain space codes. Don't try to print in lines 22 or 23, or you'll get error code 5 or error code B, depending upon how you attempted to print into them.

In the next chapter, you'll discover how and where to store machine code in memory. Later, Chapter 6 will show you how to use the display file to build a one-tenth second stopwatch.

Chapter 3

The REM(arkable) PEEK and POKE

In the last chapter you learned that information is stored in the computer at memory locations designated by addresses between 0 and 65535. Each piece of information in the TS 1000 is represented, or encoded, as an 8-bit binary number. Is there some way in which you can examine the memory locations to see precisely what is stored in them? The answer is yes, and the reasons why you'll want to do so will become clear as you proceed. We'll discuss the **PEEK** and **POKE** functions first and talk about the REM (remark) statement later in this chapter.

Most modern BASIC language dialects, including the one your TS 1000 uses, contain a **PEEK** function. As the name implies, this function allows you to peek at the contents of particular memory locations. The argument of the **PEEK** function, or the number it operates on, is the memory location whose contents you wish to examine. Because **PEEK** is a function and not a command, you must assign it to a variable or make it the argument of a command or of another function (which you've assigned to or made the argument of a command). Type

    PEEK 123

and **ENTER** it and you'll get a syntax error. This is because **PEEK** is not assigned or is not the argument of a command. Now type

    PRINT PEEK 123

Notice that your computer now responds with 253 in the upper left of the screen and 0/0 in the lower left, indicating the normal completion of a command. The 253 is the decimal value of the binary code stored at location 123 (the argument of the **PEEK** function). The **PEEK** function in this case is the argument of the **PRINT** command. Now try

 LET X = PEEK 123

and **ENTER** it. When the 0/0 appears, type

 PRINT X

Again, the 253 appears in the upper left. What you did this time was to assign the value reported by **PEEK** to the variable X and then tell the computer to print the value of X.

 You learned earlier that memory consists of up to 65536 locations, each able to store a single 8-bit number. What does this reveal about the **PEEK** function? If the biggest number that can be stored is 8 bits, then **PEEK** can never return a value larger than 255. How about negative numbers? Although conceptually possible, the **PEEK** function on the TS 1000 will interpret all values as positive. Thus, the smallest number it can report is zero. What would happen if you asked the computer to **PRINT PEEK** 123456? Try it. The B error code (integer out of range) that prints out reflects the fact that there are no memory locations above 65535 and, therefore, any integer above this value is out of range.

 Although you can now read the contents of any memory location, you're still faced with the problem of determining what the number returned by the **PEEK** function means. In general, there's no definite solution to this problem because the same binary number could be used to represent a machine code instruction, a character code, part of a 16-bit number, and so on. For instance, according to the owner's manual, the number 253 that you recovered with your **PEEK** 123 could mean the number 253, the token **CLEAR**, or a part of a machine code using the IY register. It could also represent several other things, but in this case it does mean the number 253. Unless you know beforehand what a particular location is supposed to contain, you'll find it difficult or impossible to interpret the number's meaning. An exception may occur when you have enough codes from sequential locations to determine the meaning from context.

 For now, here's a summary of what you've learned about the **PEEK** function.

- The general form for using it is ''**PEEK** argument.''
- Argument is a memory location and may have a value from 0 to 65535.

- **PEEK** is a function and must, therefore, be either
    - assigned to a variable or
    - acting as the argument of a command or acting as the argument of another function that is assigned or is itself the argument of a command.
- The value returned by **PEEK** will always be an integer between 0 and 255.
- In general, unless you know what is supposed to be stored in a specific location, you won't be able to interpret the meaning of the code returned by **PEEK**.

This last point by no means implies that **PEEK** is worthless; it simply means you must know what you're looking for before you begin. Your owner's manual lists the locations of a number of parameters you'll find useful—those ''reserved to office operations information'' and located between 16384 and 16507. This book will provide examples that illustrate how you can use these parameters to advantage; it will let your imagination find additional uses.

The system variable named RAMTOP is the address of the first memory location above 16384 that failed to record and read back the value two when the TS 1000 performed its memory test after you turned on the power. Normally, this location is the first in which RAM (*R*andom *A*ccess *M*emory) is not installed. But, if your computer has a bad RAM IC (*I*ntegrated *C*ircuit), it could be considerably less. A test (but by no means an exhaustive one) to see whether all of your memory is good is to see if the TS 1000 thinks it has as much memory as you know it has. The amount of memory that should be present is

RAMTOP – 16384

You can find out what the computer thinks this number is by typing

PRINT PEEK 16388 + 256 ∗ PEEK 16389 – 16384

You had to use **PEEK** twice because RAMTOP is a 16-bit number. The first word (at location 16388) is the low-order byte; the second (location 16389) is the high-order byte and must be multiplied by 256 because the first bit on the right of the high-order byte represents 256. (Remember, the high-order byte increments only once for each 256 times the low-order byte increments.)

FRAMES is another system variable that you may find useful. It is a 16-bit number that is decreased once for each TV frame generated. American TVs use 60 frames per second. (UK sets use 50.) If you were to look at the value of **FRAMES** at a particular time and then wait until it had changed by some specified amount, you'd have a timer. Since it changes at 60 units per second, the time in seconds is

TIME = CHANGE/60

Try this simple one-minute timer.

```
10 LET X = PEEK 16436 + 256 * PEEK 16437
20 IF PEEK 16436 + 256 * PEEK 16437 + 3600 > X THEN
   GOTO 20
30 PRINT "TIME"
```

While it will work most of the time, this timer has several problems. The most serious, which you'll repair later, is that any time X, the initial value of **FRAMES**, is less than 3600 the left-hand side of the comparison will always be greater than X, and the program will hang up in line 20. A second, less serious problem is that the execution of the BASIC program is interspersed with the generation of TV pictures so that the comparison may not occur at precisely the time **FRAMES** is different from X by exactly 3600. This is why you placed a " > " rather than an " = " in your instruction. The maximum error this discrepancy can produce is equal to the time it takes for the program, or any program that you embed this routine in, to complete one iteration; it will almost always be less than one second.

You can also get a rough idea of how much memory remains for programming by using **PEEK**. If **STKEND** is subtracted from **ERR__SP**, the resulting answer is the amount of spare memory plus the amount used by the machine stack. The latter is usually quite small, so the answer is an approximation of the amount of remaining memory. Type

PRINT PEEK 16386 + 256 * PEEK 16387 − PEEK
16412 − 256 * PEEK 16413

and **ENTER** it. Then write the one-line program

5 REM

and retype and **ENTER** the previous two lines. Notice that the number your computer returns is six less than the previous number, reflecting the six words the 5 REM statement requires. You can type this command any time you're programming to get an idea of how much memory remains for additional program lines. Remember that the actual amount is equal to or *less than* the number you obtain.

The TS 1000 and most other home computers also let you alter the contents of any memory location that's alterable, namely, any of the installed RAM locations. Because of the way it's physically built, ROM (*Read-Only Memory*) is not alterable, and an empty memory cell cannot store anything. To make alterations, you can use the **POKE** command. Its general form is

POKE arg1,arg2

Arg1 is an address between 0 and 65535, and arg2 is the *decimal equivalent* of an 8-bit binary number. The computer will accept no other numbers, symbols, or characters. You can **POKE** negative numbers, but they'll convert to two's complement notation. Your computer will report them as positive integers if you subsequently **PEEK** the **POKE**d locations. This feature is a convenience to you as a programmer because it eliminates the need for you to convert negative numbers by hand before **POKE**ing them. Because **POKE** is a command and not a function it does not have to be the argument of another command or be assigned to a variable. Instead, you can either implement it directly from the keyboard or use it as a stand-alone line in a BASIC program.

There are a number of reasons for altering the contents of specific memory locations, but the principal ones include

- Temporarily changing system parameters
- Making use of system capabilities not accessible in BASIC
- Creating faster and/or more compact programs
- Creating new files to store data, characters, or machine code

The examples in this book will explore a few of these reasons, but be aware that the reasons for **POKE**ing frequently overlap.

In the last chapter you discovered that the TS 1000 will check for installed RAM at memory locations from 16384 to 32767 and set the system variable RAMTOP to 32768 if all 16K of memory is installed. It will not, however, check above that address. If you have installed memory in the 32K–64K area, you must "tell" the computer about it. You do this by **POKE**ing into RAMTOP a value that corresponds to the address of the first memory location in which you have not installed memory. If you have added 32K, the corresponding address will be 49152, and you will

```
POKE 16389,192
POKE 16388,0
```

If you added 48K, 56K, or 64K memory packs, then the address should be 65536. In binary the number is a one followed by 16 zeros and cannot be **POKE**d into RAMTOP, itself a 16-bit number. Instead, use the address 65535 and

```
POKE 16389,255
POKE 16388,255
```

Your computer now thinks location 65535 is empty and inaccessible to the BASIC system (even though you know a memory cell is there). This is insignificant unless you write a BASIC program that requires exactly 49152 storage locations—a very unlikely event. Note that for memories larger than 48K there is no way to tell the computer that the 8K–16K area is populated. This area is forever out of the reach of the BASIC system except via **PEEK** and **POKE**.

Now you can see how the **POKE** command can be used to solve the hang-up problem you had earlier with the simple timer. Recall that the problem revolved around the fact that the system variable **FRAMES** could contain a value less than the number of frames you wanted it to count down (3600 in the example). This problem can be rectified by deliberately **POKE**ing **FRAMES** with the largest number it can contain each time you want to exercise the timer. Type the following program.

```
10 POKE 16437,255
20 POKE 16436,255
30 IF PEEK 16436+256*PEEK 16437>61935 THEN
   GOTO 30
40 PRINT ''TIME''
```

What you have done is to **POKE FRAMES** with the value 65535 and then **PEEK** at it until the value has decremented at least 3600 (65535−3600=61935) at which time the TS 1000 will print ''TIME'' on the screen. The operating system of the computer does the decrementing automatically; it requires no BASIC control. For any other interval T, in seconds up to 1080 (18 minutes), substitute the value of N for 61935.

$$N = 65535 - T*60$$

Compared with the **PAUSE** command, which also uses **FRAMES**, this method of timing has the advantage of not causing screen flicker and allowing other program steps to take place during the interval. (For example, you could call for tests, operations, and printing between lines 20 and 30.) Further, if you are content to time in 4.25-second intervals with up to a 4.25-second error, you can reduce the program to

```
10 POKE 16437,255
20 IF PEEK 16437 > N THEN GOTO 20
30 PRINT ''TIME''
```

In this case

$$N = 255 - INT (.5 + T/4.25)$$

Here, you're dealing with only the high-order byte of **FRAMES**, which will decrement once every 4.25 seconds. The modified program is suitable for coarse timing and requires about half the memory space of the longer version, a prime consideration if you have a 1K memory.

Direct manipulation of the display file by using the **POKE** command is a technique that you can frequently use to speed up graphics routines. This is because **POKE** is a faster command than either **PRINT** or **PRINT AT**. (See Chapter 6.) It involves determining the address in memory of the location that corresponds to a particular row and column on the screen and then **POKE**ing the code for the character of your choice into that location.

In the previous chapter you learned the general structure of the display file, which contains the echo of the TV screen. The echo starts at the address that the system variable D__FILE provides; its first code is that of the **ENTER** character. Next, each line on the screen is stored sequentially, starting with the top line. Each ends with an **ENTER** character. There are 25 **ENTER**s in all, one for the start-of-file character and one to end each of the 24 screen lines.

The amount of data stored in each line depends on how much memory you have. If you have more than 3.25K, then the display is padded out with 24 lines of 32 spaces, meaning that each line in memory will consist of 32 space codes and an **ENTER** code. If you have less memory, the display file is collapsed; it contains only the 25 **ENTER** codes until you actually print something. Once you print something, the memory will contain the codes for the printed characters in the locations that correspond to the line in which they were printed plus an **ENTER** code ending the line. (This is in addition to the other 24 **ENTER**s.) An exception occurs with **PRINT AT**. Here, the beginning of the line in which the **PRINT AT** occurred is padded out with spaces to get to the column at which the **PRINT AT** occurred.

Assume you have only the 2K memory and pad out part of the display file. (The example works just as well if you have additional memory.) Make an asterisk flash 1000 times in line 5, column 15 using the program that follows.

```
10 FOR I = 1 TO 1000
20 PRINT AT 5,15;"*";AT 5,15;"  "
30 NEXT I
```

Time how long it takes the asterisk to complete the 1000 flashes (you should count about 68 seconds), and then type the following program.

```
10 PRINT AT 5,15;"  "
20 LET X = PEEK 16396 + 256 * PEEK 16397 + 20
30 FOR I = 1 TO 1000
30 FOR I = 1 TO 1000
40 POKE X,23
50 POKE X,0
60 NEXT I
```

Time how long the 1000 flashes take using this program. Did you get 52 seconds? Using **POKE** to directly change the display file resulted in the graphic display taking only 3/4 as long. Such time savings can be important when you're creating games and other interactive displays. You'll see more of this technique in Chapter 6.

One word of caution: *do not* try to **POKE** a collapsed display file. Your system will crash. Your computer will hang up in never-never land, and you will have to do a power-on reset to get it back, destroying everything you've already programmed. This is because the operating system of the TS 1000 must see 25 **ENTER** characters in the display file in order to function properly, and if you **POKE** the empty file, you'll have overwritten one or more of those characters. This is why you printed a space at line 5, column 15 in the second program before you **POKE**ed the asterisk code.

Now for the creation of files for your own use. For your purposes, a file constitutes nothing more than a section of memory blocked off for a particular use. Such files might contain numerical data, character strings, or machine codes. You could, for instance, designate a section of spare memory as a ''machine code'' file and stick all of the routines you'll develop in it. However, in doing so you run the serious risk of having the existing files overwrite it as your BASIC programs, displays, variables, and other files grow in length. (Remember, each time you type a new program line, print a character, or assign a variable, the address of the lower edge of spare memory increases towards RAMTOP to make room for the new material. With enough additions, you'll end up overwriting your spare memory, including the portion you designate for your own use.) Where, then, do you store your files? Your owner's manual suggests three places.

- In a string
- In a REMark statement
- Above RAMTOP

If you have more than 48K of memory, then the 8K–16K area of memory is also available.

Compared with storing data in REM statements, storing data in strings offers no advantages except in the case of character strings, where the BASIC system automatically keeps track of the string variables for you. In fact, string storage has the significant disadvantage of not allowing absolute, or immediate, addressing. That is, because the variables area slides as program length changes, you can't tell your computer ''Look at address 17203'' and be sure that the information you desire is at that location. Rather, you must use a combination of indirect and relative addressing. If the piece of data you need is stored at the tenth location in the variables area, you must say things like ''Look at the address given by the system variable **VARS** (indirect) plus ten (relative).'' For this reason you won't be making use of variable storage in any of the projects this book describes.

Memory locations above RAMTOP, or in the 8K–16K area, have the distinct advantage of being inviolate to keyboard commands. Even the **NEW** command will not destroy their contents. In addition, if you use memory above 32K or between 8K and 16K, and you've installed a RESET button (see Chapter 8), then you can recycle the TS 1000 when it hangs up without destroying what you've stored at those locations. This is because the button, like turning the power off and on, branches the computer back to the beginning of the BASIC system, destroying anything in the 16K–32K block. (The memory tests at power-on.) But unlike turning the power off and on, using RESET doesn't erase the additional memory, anything stored there remains intact. Therefore, memory locations above RAMTOP (and above 32K in particular) or those in the 8K–16K section are ideally suited for storing utilities.

Utilities are routines that usually alter or add to system behavior and are useful with a variety of programs. For example, routines that add sound, scroll the screen down instead of up, delete blocks of BASIC lines, renumber programs, and so forth may all be considered utilities, and you might want to retain them for use with any program you're writing. The disadvantage of these locations is that their contents are not written to tape when you **SAVE** a program. To save them you must either write a machine code save routine or transfer the codes in some acceptable manner to locations that will be **SAVE**d. An additional disadvantage of the 8K–16K block is that many manufacturers of peripheral devices for the TS 1000 use this area for memory mapped I/O. If you have attached such a device to your computer, you may not be able to use this block. As a partial solution you can use peripherals that occupy either the 8K–12K or 12K–16K areas and memory packs that allow switch selection of which of the two areas you'll use for memory.

Now for the REM statement. As you discovered in Chapter 2, if a REM is the first line of a BASIC program, then the argument of the REM always begins in location 16514. Furthermore, REMs are saved by the **SAVE** command. Thus, if you store your data as the argument of a REM, you can use immediate mode addressing, and your data will be written to tape whenever you **SAVE** your program. The disadvantages are that REMs

- Will be wiped out by the **NEW** command or a power-on reset.
- Eat up potential BASIC program memory.
- Cannot contain the ENTER code 118. (This is because all BASIC lines end with ENTER, and one contained within the line would erroneously signify the end of the line.)

Because of their recordability, REMs make a good place to store information. And, if you have 16K memory or less, the BASIC memory penalty associated with using REMs instead of the area above RAMTOP is only six words. Feel free to try using other areas, although the projects in this book will use REMs almost exclusively.

You can perform an experiment to see how REMs hold data. Suppose you wish to store the decimal integers 255, 0, 73, and 58. Type this one-line program.

    10 REM 1234567890

Notice that it echoed on the upper part of the screen just as you typed it. Now enter the following **POKE**s.

    POKE 16514,255
    POKE 16515,0
    POKE 16516,73
    POKE 16517,58

Where did that "COPY ?U" come from? If you'll look at the listing of character codes in your owner's manual, you'll see that the numbers you **POKE**d into the argument of the REM statement are the codes for COPY, space, ? (? occurs for any of the unused codes), and U. Thus, you've not only successfully **POKE**d the numbers into known locations but you've also automatically received a confirmation that the computer stored the proper numbers. Because the original 1–4 in line 10 were overwritten but 5–0 remain, you also know that you **POKE**d four numbers.

If you're sharp, you're probably wondering why you didn't just type the REM with those characters already in it instead of **POKE**ing them. How would you have typed in that question mark? You couldn't have used the keyboard question mark—that's a 15 code. There are a number of other untypeable codes, so for consistency and clarity you'll continue to **POKE** codes rather than type characters. If you wish to check the stored numerical values directly, type

    PRINT PEEK 16514, PEEK 16515, PEEK 16516, PEEK 16517

With this technique you can sequentially store any number of 8-bit numbers. By grabbing 2, 3, 4 words at a time you can also store 16-, 24-, 32-bit numbers. In each case, however, you have to increase the length of the argument of the REM statement to accommodate the total number of words you want to store. It may be most convenient to do this by repeating the 1–0 sequence as many times as necessary. The total available storage is then the number of zeros times ten. If you're into hex (hexadecimal, base sixteen notation), you might want to use a 0–F sequence and multiply the number of Fs by 16—or use the alphabet and multiply the number of Zs by 26. In every case, if you're not too pressed for program memory space, you can use a longer argument than necessary and either ignore the extra characters or delete them when you've finished storing your data. To avoid typing all of the characters in the argument of the REM, the rest of the book will use notation like

    10 REM (23 characters)

This means that the number of characters following the REM must be at least 23. They can be any character of your choice. It *does not* mean you should type ''(23 characters).''

So far, so good. But what if you want to save 50 numbers, and not just a number or two. You don't want to do all that **POKE**ing and **PEEK**ing—nobody does. That's why you should use Programs 3.1A and 3.1B. They're called a machine code loader and reader for reasons that will become clear in the next chapter, but they are, in fact, just simple BASIC routines for sequentially **POKE**ing or **PEEK**ing a series of memory locations. They both start at 16514 so that you can use REMs for storage, but if you wish to store data elsewhere, then substitute the starting address of your preferred location for the first number in lines 8000 and 9000. The ending location for I is arbitrary; it must be sufficiently larger than the starting address to include all of the locations you wish to use. You will also have to change the number in lines 8040 and 9040 to equal your starting address plus 20. Or, if you are going to **POKE** or **PEEK** fewer than 23 items, you can delete 8040 and 9040 entirely. Trying to enter anything but a decimal integer between 0 and 255 inclusive will halt the loader, so simply type a letter when you have entered everything that you desire. When the reader is running, your hitting any key will cause it to pause for about nine minutes. This is important only when your computer is trying to read more than 22 locations. Hitting any key during the pause will cause the program to resume. For 22 or fewer items, you can delete line 8030. The program will terminate itself it it encounters an **ENTER** code. Otherwise you can stop it by hitting **BREAK**.

Now that you know where and how to store decimal integers in the TS 1000 you should try using the machine code loader and reader a time or two with a 50-element REM statement to get the feel of them. Try **POKE**ing the first and then the second 50 integers into the REM. Do the characters that result agree with those your owner's manual's code table? Does the reader bring

back the proper values? When you're comfortable with these
operations, go on to the next chapter.

```
8000 FOR I = 16514 TO 32000
8010 INPUT A$
8020 POKE I, VAL A$
8030 PRINT I, PEEK I
8040 IF I > 16534 THEN SCROLL
8050 NEXT I
```

PROGRAM 3.1A Machine Code Loader

```
9000 FOR I = 16514 TO 32000
9010 IF PEEK I = 118 THEN STOP
9020 PRINT I, PEEK I
9030 IF INKEY$ < > "" THEN PAUSE 32000
9040 IF I > 16534 THEN SCROLL
9050 NEXT I
```

PROGRAM 3.1B Machine Code Reader

Chapter 4

Learning Another Language

In the last chapter you discovered how to **POKE** numbers into various locations in the TS 1000 memory. In this chapter you'll learn that you can use those numbers to encode sequences of machine instructions that can cause the computer to do things not possible with the BASIC operating system. The experience is much like learning a new language, but with an important difference: the "rules of grammar" and the "spelling" in your new language are extremely rigid. With a natural language like English, you can occasionally misspell a word, use a wrong tense, or drop a participle, and still convey much of your original meaning. In machine coding, any error in any numeral can completely foul up the program you are trying to develop. This need for precision is both a curse and a blessing: it requires that you be meticulous in entering codes, but it also eliminates any ambiguity in definitions or command string interpretation.

Every microprocessor (mP) or computer has its own machine language, but in many cases these languages are as similar as Spanish is to Portuguese or as American English is to British English. In other cases they differ more significantly. The language you're about to begin learning is that of the Z80 microprocessor, and you can use the instructions you'll learn on any Z80-based computer. (You may not be able to use all the programs; they also depend on the hardware attached to the Z80 in a particular computer.) The individual instructions available to the mP may be considered the vocabulary of that device. Like natural (human) languages, the total vocabulary of a machine language may be quite large (the Z80's contains about 500 words, making it one of the more powerful instruction sets on any micro), but it's not necessary to know all the words to communicate effectively. Many of the words may be considered "scholarly" types reserved for those with advanced programming experience. Still others (such as the **BIT**, **SET**, and **RES** commands of which there are close to 200—and which you will not use in these projects) are simply variations of the same fundamental commands.

This chapter will expose you to just a small subset of Z80 instructions, but the 35 it shows you will be sufficient for any of the programs you'll develop later in the book. What's more, if you can master these, you'll probably have little trouble expanding your vocabulary at a later time, given a good text like *Programming the Z80* by Rodnay Zaks and *Z80 User's Manual* by Joseph J. Carr. Both are excellent books but can be a bit heavy going unless you've had some experience with machine coding.

Strictly speaking, a machine-coded program for the Z80 would consist of a series of 8-bit binary numbers, since this is the only thing the hardware can understand. Doubtless you would find this a difficult and tedious method of programming. And, as you discovered in the last chapter, you'd be unable to **POKE** these numbers into the TS 1000. What you'll do instead is use the *decimal equivalent* of the 8-bit codes. Your computer will automatically convert them to binary numbers in memory. As an aside, you'll find that the authors of many other books, with considerable justification, use the hexadecimal (hex) equivalents of the machine codes to describe their programs. But since the TS 1000 does not accept hex, this book will refer only to the decimal equivalents. "Fine," you say, "but a string of decimal integers is just barely better than a string of binaries." Right you are! For this reason the program steps will express the machine language instructions in their assembly language mnemonics. These mnemonics are short words or letter groupings that, by their sound or spelling, signify what the operation is supposed to do. Some typical examples include

DEC B   Decrease the B register.

IN A, 254   Input to the accumulator from port 254.

LD B, H   Load the B register with the contents of the H register.

Not so hard to understand, is it? Table 4.1 is a complete alphabetical list of the instructions you'll need. Wherever possible, the book will point out the similarities between the machine instructions and BASIC commands. Be sure to study it carefully before continuing; you may need to refer to it frequently throughout the rest of the book.

TABLE 4.1 Subset of Z80 Machine Instructions

ADD HL, DE   CODE 25

One-word instruction. Sixteen-bit addition. Add the contents of the DE register pair to the contents of the HL register pair and store the result in HL. Carry flag is set by a carry from bit 15. Zero flag is unaffected. Requires 11 clock cycles. BASIC: **LET** HL = HL + DE

AND C   CODE 161

One-word instruction. Logically AND the contents of the C register with those of the accumulator. Store the result in the accumulator. Carry flag is reset. Zero flag is set or reset according to the outcome of the operation. Requires 4 clock cycles. No direct BASIC equivalent; not the same as BASIC AND.

AND n   CODES 230, n

Two-word instruction. Second word is number n. Logically AND the 8-bit number n with the contents of the accumulator. Store the result in the accumulator. Carry flag is reset. Zero flag is set or reset according to the outcome of the operation. Requires 7 clock cycles. No direct BASIC equivalent; not the same as BASIC AND.

CALL qp   CODES 205, p, q

Three-word instruction. Call (branch to) the subroutine starting at address qp. Second and third words are low-order and high-order bytes, respectively, of the starting address of the subroutine. Flags are unaffected. Requires 17 clock cycles. BASIC: **GOSUB** xxxx

CP n   CODES 254, n

Two-word instruction. Second word is number n. Compare the number n with the contents of the accumulator. (Note that n is subtracted from the accumulator, and the result discarded.) Both the carry and zero flags are set or reset according to the outcome of the operation. Requires 7 clock cycles. BASIC: **IF** A = n **THEN** (reset zero)

CPL   CODE 47

One-word instruction. One's complement the accumulator and store the complemented result back in the accumulator. Carry and zero flags are unaffected. Requires 4 clock cycles. No direct BASIC equivalent.

DEC C   CODE 13

One-word instruction. Decrement (decrease by one) the contents of the C register. Carry flag is unaffected, and zero flag is set or reset according to the outcome of the operation. Requires 4 clock cycles. BASIC: **LET** $C = C - 1$

DJNZ e   CODES 16, e − 2

Two-word instruction. Decrement the B register and, if the result is not zero, then jump $e - 2$ locations. Suppose this instruction is at address qp and you wish to jump e locations. At the end of this instruction, the PC register will contain the address of the next sequential instruction, which should be $qp + 2$. Therefore, the displacement, e, is given an offset of −2 to compensate. The displacement itself is in two's complement so that the jump range will be effectively −126 to +129. You can jump forward or backward in the program. Remember that a backward jump is negative and that adding −2 to a negative results in a negative; i.e., $-10 - 2 = -12$. Jumps larger than this are not permissible with this instruction. Flags are not affected. Requires 13 clock cycles if B is not zero and 8 clock cycles if B is zero. BASIC: **LET** $B = B - 1$, IF $B < > 0$ **THEN GOTO** current line + e lines

IN A, 254   CODES 219, 254

Two-word instruction. Input to the accumulator from input port 254. The number 254 goes into the low-order byte of the address bus (A0–A7), and whatever was in the accumulator goes into the high-order byte. The inputted value ends up in the accumulator. The input port is a piece of hardware that will respond to 254 on the mP address bus and to IORQ and RD signals on the control bus. Because of hardware considerations in the TS 1000, port 254 is the only input port you'll access. Flags are unaffected. Requires 11 clock cycles. No BASIC equivalent.

INC BC   CODE 3

One-word instruction. Increment (increase by one) the contents of the BC register pair. Flags are unaffected. Requires 6 clock cycles. BASIC: **LET** $BC = BC + 1$

INC HL   CODE 35
>One-word instruction. Increment the contents of the HL register pair. Flags are unaffected. Requires 6 clock cycles. Same as previous instruction but for HL rather than BC. BASIC: **LET** HL = HL + 1

JR C, e   CODES 56, e − 2
>Two-word instruction. This is a conditional branch instruction; jump e locations relative to current location if a carry flag is set. The relative displacement, e, works in exactly the same fashion it did with DJNZ, allowing conditional jumps of −126 to +129. Requires 12 clock cycles if carry is set and 7 if carry is not set. BASIC: **IF** CARRY = 1 **THEN GOTO** current line + e lines

JR e   CODES 24, e − 2
>Two-word instruction. This is an unconditional branch instruction. Program will jump e locations relative to current location whenever this command is encountered. Relative displacement, e, same as for DJNZ, allowing relative jumps of −126 to +129. Requires 12 clock cycles. BASIC: **GOTO** current line + e lines

JR NC, e   CODES 48, e − 2
>Two-word instruction. Exactly the same as JR C, e, except the condition is now jump if there is no carry (carry is reset). Requires 12 clock cycles if carry is reset and 7 if carry is set. BASIC: **IF** CARRY = 0 **THEN GOTO** current line + e lines

JR NZ, e   CODES 32, e − 2
>Two-word instruction. Exactly the same as JR C, e, except the condition is now jump if not zero (zero flag reset). Requires 12 clock cycles if zero is reset and 7 if zero is set. BASIC: **IF** ZERO = 0 **THEN GOTO** current line + e lines

JR Z, e   CODES 40, e − 2
>Two-word instruction. Exactly the same as JR C, e, except the condition is now jump if zero (zero flag is set). Requires 12 clock cycles if zero is set and 7 if zero is not set. BASIC: **IF** ZERO = 1 **THEN GOTO** current line + e lines

LD A, n   CODES 62, n
>Two-word instruction. Load the accumulator with the number n. The 8-bit number n is placed in the A register or the accumulator. Flags are unaffected. Requires 7 clock cycles. BASIC: **LET** A = n

LD B, n   CODES 6, n
> Two-word instruction. Load the B register with the number n. The 8-bit number n is placed in the B register. Flags are unaffected. Requires 7 clock cycles. BASIC: **LET** B = n

LD C, n   CODES 14, n
> Two-word instruction. Load the C register with the number n. The 8-bit number n is placed in the C register. Flags are unaffected. Requires 7 clock cycles. BASIC: **LET** C = n

LD E, n   CODES 30, n
> Two-word instruction. Load E register with the number n. The 8-bit number n is placed in the E register. Flags are unaffected. Requires 7 clock cycles. BASIC: **LET** E = n

LD B, H   CODE 68
> One-word instruction. Load the B register with the contents of the H register. The 8-bit number contained in H is placed in B. Both registers now contain that number. Flags are unaffected. Requires 4 clock cycles. BASIC: **LET** B = H

LD C, A   CODE 79
> One-word instruction. Load the C register with the contents of the A register. The 8-bit number contained in A is placed in C. Both registers now contain that number. Flags are unaffected. Requires 4 clock cycles. BASIC: **LET** C = A

LD C, L   CODE 77
> One-word instruction. Load the C register with the contents of the L register. The 8-bit number contained in L is placed in C. Both registers now contain that number. Flags are unaffected. Requires 4 clock cycles. BASIC: **LET** C = L

LD DE, mn   CODES 17, n, m
> Three-word instruction. Load the DE register pair with the 16-bit number mn, low-order byte first. This could also be stated as load D with m and E with n. In either case, the 16-bit number mn is placed in the DE register pair. Flags are unaffected. Requires 10 clock cycles. BASIC: **LET** DE = mn

LD HL, mn   CODES 33, n, m
> Three-word instruction. Load the HL register pair with the 16-bit number mn, low-order byte first. This could also be stated as load H with m and L with n. In either case, the 16-bit number mn is placed in the HL register pair. Flags are unaffected. Requires 10 clock cycles. BASIC: **LET** HL = mn

LD (HL), A   CODE 119

One-word instruction. Load the address given by the contents of the HL register pair with the contents of the accumulator. HL contains a 16-bit number considered an address in memory into which the number contained in the A register is to be placed. This is known as indirect addressing. Flags are unaffected. Requires 7 clock cycles. No direct equivalent in BASIC, but sometimes **POKE** and **PEEK** can be used in a similar manner: **POKE PEEK** (HL), A

NOP   CODE 0

One-word instruction. NOP means no operation. The mP does nothing for 4 clock cycles and then passes on to the next sequential instruction. Useful for timing. Flags are unaffected. Requires 4 clock cycles. No equivalent in BASIC.

OUT 255, A   CODES 211, 255

Two-word instruction. The contents of the accumulator are sent to output port 255. The output port is a piece of hardware that will respond to an address of 255 on the mP address bus and to IORQ and WR signals on the control bus. Because of hardware considerations within the TS 1000, you'll access only output port 255. Flags are unaffected. Requires 11 clock cycles. No BASIC equivalent.

RET   CODE 201

One-word instruction. Return from subroutine. Used either to return from a machine code subroutine to a machine code calling routine or to return from a machine code routine to BASIC control. You'll use it in the latter sense. The machine code routines you'll develop may be considered subroutines of our BASIC programs so that RET will function in exactly the same way as the BASIC ''RETURN.'' Flags are unaffected. Requires 10 clock cycles.

RET C   CODE 216

One-word instruction. A conditional return executes only if carry is set; otherwise, the same as RET. Control passes to next sequential instruction if carry is reset. Requires 11 clock cycles if carry is set and 5 if carry is not set. BASIC: **IF CARRY = 1 THEN RETURN**

RET NC   CODE 208

One-word instruction. A conditional return executes only if carry is reset; otherwise, the same as RET. Control passes to next sequential instruction if carry is set. Requires 11 clock cycles if carry is reset and 5 if carry is not set. BASIC **IF** CARRY = 0 **THEN RETURN**

RLA   CODE 23

One-word instruction. Rotate the accumulator left through the carry bit of the flag's register. (Imagine nine boxes all in a row. The 8 bits of the A register are in the eight boxes on the right, with the lowest order bit on the extreme right and the highest order bit in the eighth box from the right. The carry bit is in the ninth box, that on the extreme left. Shift each bit one box to the left. Take the bit that popped out of the carry bit box, bring it back around to the extreme right, and place it in that box.)

	C	A Register
Before	1	01110101
After	0	11101011

The carry bit is obviously set or reset according to the outcome of the operation. The zero bit is unaffected. Requires 4 clock cycles. No BASIC equivalent.

RRA   CODE 31

One-word instruction. Rotate the accumulator right, through the carry bit of the flags register. The same as RLA except that the direction of rotation is to the right rather than to the left. (If you took the ''After'' example in RLA and performed an RRA, you would end up with the original ''Before.'') Carry flag is set or reset according to the outcome of the operation. Zero flag is unaffected. Requires 4 clock cycles. No BASIC equivalent.

RR C   CODES 203, 25

Two-word instruction. Rotate the C register right through the carry bit of the flag's register. Performs the same function as RRA but uses the C register instead of the accumulator. The carry bit will be set or reset according to the outcome of the operation. Zero bit is unaffected. Because this command is one of a set of rotation instructions more flexible than RRA, it requires two words and takes 8 clock cycles. No BASIC equivalent.

What is a machine code program? Like a BASIC program, it's a sequence of instructions that the computer will follow to complete the task you designed the program to handle. Normally, your computer will carry out these instructions in the sequence in which you wrote them, just as a BASIC program proceeds from the lowest numbered line to the highest. However, just as in BASIC, machine code programs may contain branch instructions. These direct the mP to carry out some instruction other than the next in line. In BASIC, these instructions are the **GOTO**, **GOSUB**, and **IF . . . THEN GOTO** commands. The last is called a conditional branch, since the route of program execution will change only if the particular condition is satisfied; otherwise, the next sequential instruction will execute. The **JP**, **CALL**, and **JR c/JP c** family of commands serves the same purpose in machine coding. A significant difference is that each machine code instruction, or "line," requires a known, unvarying amount of memory space, usually one to three words depending on the instruction. BASIC lines vary in length. This means that you don't have to number machine code program lines: you can find your way around the program by counting memory locations. Naturally, the sequence must start in a known memory location. You'll be storing all your programs in a REM statement, which is the first line of a BASIC program. So, as you found in the last chapter, the starting location will always be 16514. Of course, you may use other areas if you desire.

Before you can write a program, you must design it. Design techniques are essentially the same for machine and BASIC programming. (You'll find that as you become more familiar with machine codes you'll be able to write short routines right off the top of your head, just as you've probably learned to do in BASIC. But for now, stick to the formalities—it'll help you to avoid problems while you're learning.)

First, state the problem as carefully as possible. Then, list the major steps in solving it on the computer. It's difficult to over-emphasize the importance of these initial steps. There are almost always several ways to solve the same problem, but most of them will involve extra steps or they'll use instructions that are inherently more time consuming (require more memory, include awkward logic, are slower to execute). Careful thought at the beginning can save you untold frustration later. After you've had the experience of working on a hastily constructed routine for several hours only to scrap it because of fundamental errors in logic, you may appreciate this point!

From the list of steps you can create a primitive flowchart. (See your owner's manual for a short tutorial on flowcharting.) If the routine you're developing is relatively simple, you may be able to go directly from this flowchart to the program. More often, you'll want to expand the original chart by breaking down the major steps into smaller units. Eventually, you could expand your flowchart to the point where each block corresponded to a machine instruction, but this is overkill. Experience will quickly show you the proper level of expansion. Most likely, this level will have each block on the chart corresponding to a short subroutine or a discrete section of the program. In the example that follows shortly, you'll use flowcharts; in the remaining chapters you won't. But they played a part in the development of programs for this book.

With the flowchart complete, you can begin writing the program. You may find it easiest to write the list of mnemonics corresponding to the steps on the chart. Work on one step at a time. When you've satisfied yourself that the resulting routine will probably execute properly, continue with the next step until all the steps are complete. If you're using branch instructions, label the points that the program is to jump to, since you don't as yet know what the displacement will be.

```
LABEL       STATEMENT TO BE JUMPED TO
              ⋮
            MORE STATEMENTS
              ⋮
            JR C, LABEL
```

Later you'll count the steps to be jumped and substitute them for LABEL in the JR C command.

You must now translate your list of statements into **POKE**able codes and determine where they are to be **POKE**d. This requires little more than careful bookkeeping. You could use a "ledger" like this one.

Location	Code	Comments
16514	33	LD HL
16515	0	low-order byte
16516	1	high-order byte (256)
16517	30	LD E
16518	11	value of E = 11
16519	68	LD B, H
16520	77	LD C, L
16521	201	RET (return to BASIC)

It lists a program for loading the HL register pair with the value 256, the E register with 11, and for transferring the values in HL to BC. From Table 4.1 (already shown) you see that LD HL, 256 is a three-word instruction, low-order byte first; 256 in binary is 00000001 00000000; the low-order byte is 0, and the high-order byte is 1. Consequently, the first three locations contain, sequentially, 33, 0, and 1. Similarly, LD E, 11 is a two-word instruction, so the next two locations contain 30 and 11. The next two locations contain the 68 and 77 for LD B, H and LD C, L. The RET in 16521 is necessary if you want control to return to the BASIC system after the machine code routine is complete. Leaving it off will probably result in a system crash.

The problems with this format are that the mnemonics are split up into two or more lines, and you're writing extra lines just to list the data bytes. This is perhaps acceptable while first learning machine code, because adding such detail makes your program very specific with respect to storage locations. Most people prefer the following form, and you'll be using it, for the most part, throughout the remainder of the book.

Starting Location	Codes	Mnemonic
16514	33, 0, 1	LD HL, 256
16517	30, 11	LD E, 11
16519	68	LD B, H
16520	77	LD C, L
16521	201	RET

This form is more compact, it keeps the mnemonics together, and it leaves associated codes on the same line. Note, however, that the locations in the left hand column are no longer sequential; they are the starting locations for the statement in that line of the program. If a line contains a three-word instruction, then the next location will be three higher than the last.

Having come this far, you need only **POKE** your codes into memory. Since this program contains eight codes, you'll need a REM statement with at least eight characters. Type

```
10 REM 12345678
```

and then simply **POKE** the codes into memory sequentially starting at 16514.

```
POKE 16514,33
POKE 16515,0
POKE 16516,1
POKE 16517,30
POKE 16518,11
POKE 16519,68
POKE 16520,77
POKE 16521,201
```

You could, of course, have used the machine code loader from the last chapter, but with a program this short it is easier to directly **POKE** the codes than to key in the loader.

Since the BASIC system ignores REM statements at the time of program execution, how do you get your routine to run? Mr. Sinclair was thoughtful enough to provide you with the USR function. It has the general form

USR arg

where arg is an address between 0 and 65535. It instructs the computer to go to the address given by the argument and execute whatever code is found there. The computer will continue executing codes until it encounters a RET command, at which time it will return to BASIC. **USR** is a function, though, and so must be assigned to a variable or it must itself be the argument of a command. In the first case it will have the general form

LET X = USR arg

When used in this fashion, the value contained in the BC register pair at the completion of the machine code routine will be returned to BASIC as the value of X. Add these two lines after your REM statement.

20 LET X = USR 16514

30 PRINT X

Run this program. Did your computer print 256? If not, you've entered something wrong. The program places 256 in the HL pair and then transfers the value to the BC pair. At the completion of the routine the value in BC will be assigned to X, and the computer will print it. Assignment takes up both time and memory. If you have a routine that is supposed to do something, as opposed to give you answers, then it's better to make **USR** the argument of a command. The argument can be any command, but many will do things within the parent BASIC program you don't want. One command you can almost always use with impunity is **RAND**. The general form is

RAND USR arg

If you delete lines 20 and 30 and type **RAND USR** 16514 in immediate mode, your machine program will still run, but you'll have no way of knowing about it. In the rest of the book you'll usually use the **RAND** construct, since none of these projects require the machine code routines to return answers to BASIC. But what would happen if you had to return more than one answer? A convenient way to handle this situation would be to have the machine code routine store the numbers in particular memory locations and then **PEEK** those locations from BASIC.

As a review of the material covered in this chapter, try writing a short routine to add two 16-bit numbers, to print the answer if it is correct or signify that the answer is wrong if overflow occured. (Recall from Chapter 1 that overflow occurs whenever the sum of the two numbers exceeds 65535.) Having read the problem, begin by listing the major steps in solving it.

Load the numbers.
Add them.
Test the result.
Too large? Print error message.
Small enough? Print answer.

From these steps, construct your first flowchart, as in Figure 4.1. Next, expand the flowchart. Loading numbers presents no problem; from the table of instructions you'll see that you can directly load 16-bit numbers into the HL and DE register pairs. Similarly, ADD HL, DE lets you directly add the numbers and store the result in HL. Suppose you want to do the printing in BASIC. You must then transfer the answer to the BC register pair so it will be returned to the BASIC system. LD B, H and LD C, L allow you to do this easily. But how do you find out if the result is too large? Any time the ADD HL, DE instruction is used and the result is greater than 65535, the carry bit of the flags register will be set. Note that loading the BC pair with the contents of the HL pair does not change the status of the carry flag. Therefore, if the carry is set, you know the answer is too large and your computer should signify that an error has been made.

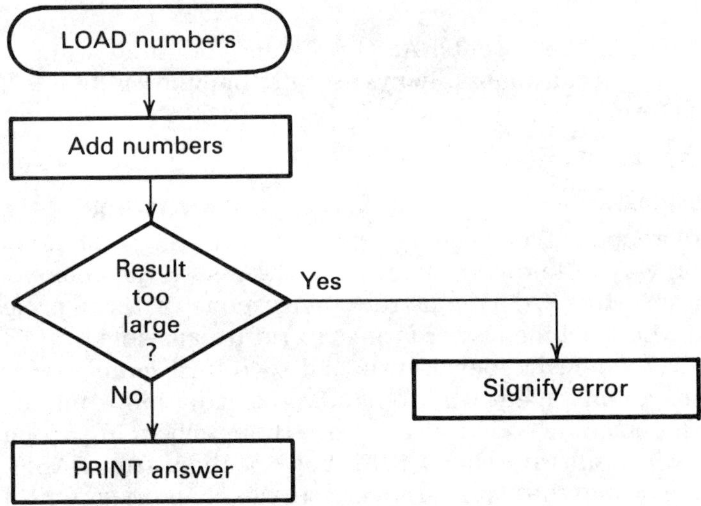

FIGURE 4.1 Simple Flowchart for Adding Two 16-Bit Numbers

Make the TS 1000 preceed the answer with the letter B (error code for integer out of range) when overflow occurs. The expanded flowchart now looks something like that in Figure 4.2. As you can see, the flowchart almost expresses the whole program. The one remaining problem is how to print ''B.'' You can't use the BC pair to return the code for the letter B because you're already using it to return the answer. One approach would be to store the code for B in some other location in memory and then **PEEK** that location from BASIC. Fortunately for you, there's a more clever approach, and you can execute it entirely in machine code. In your computer's ROM (*Read-Only Memory*) at location 16 there's a routine that prints the character whose code is contained in the A register. All you have to do is load A with the code for the letter B and then call that routine as a subroutine of our program.

So now list the mnemonics for your program.

```
          LD HL, first number
          LD DE, second number
          ADD HL, DE
          LD B, H
          LD C, L
          JR C, LAB1
          RET
LAB1      LD A, code ''B''
          CALL TS 1000 print-a-character routine
          RET
```

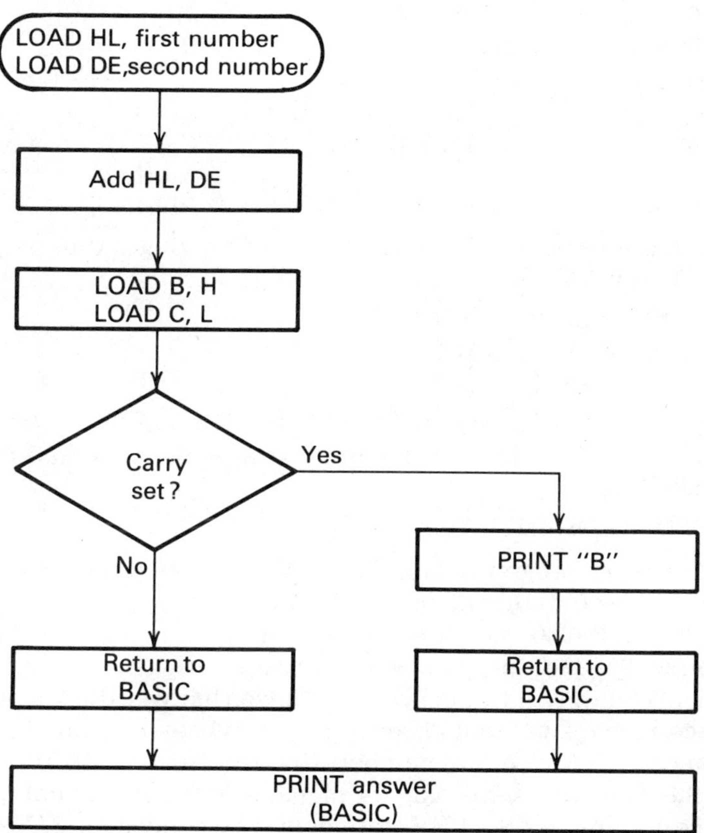

FIGURE 4.2 Expanded Flowchart for Adding Two 16-Bit Numbers

Notice that if carry is reset, you simply return to BASIC with the
answer in the BC pair, whereas if it is set, you jump down and load
the accumulator with the code for B, print it, and then return to
BASIC with the (overflowed) answer still in the BC pair and ready
to be printed in BASIC. All that remains is for you to encode these
statements.

Starting Location	Codes	Mnemonic
16514	33, 0, 1	LD HL, 256
16517	17, 0, 130	LD DE, 33280
16520	25	ADD HL, DE
16521	68	LD B, H
16522	77	LD C, L

16523	56, 1	JR C, 3
16525	201	RET
16526	62, 39	LD A, 39 (code for "B")
16528	205, 16, 0	CALL 16 (print routine)
16531	201	RET

POKE these codes into the REM in the following BASIC program and **RUN** it. Use the machine code loader.

```
10 REM 123456789012345678
20 LET X = USR 16514
30 PRINT X
```

The numbers originally in HL and DE, 256 and 33280, respectively, will cause the sum, 33536, to print in the upper left of the screen. Now

```
POKE 16516,131
```

and **RUN** the program again. The answer, 1280, is now preceded by the letter B. Is that number meaningless? Not at all. The sum of 33536 and 33280, the numbers placed in HL and DE, is 66816. Subtract 66536 from it. Did you get 1280? With this approach you not only know that the answer overflowed but you also know that the correct answer should be 65536 plus whatever follows the B. (You can effectively add numbers that total up to 131070.) To do all this you have used only 24 memory locations including the space required for the REM statement. The equivalent BASIC

```
10 LET X = first number
20 LET Y = second number
30 LET X = X + Y
```

would require upwards of 40 locations and would not execute nearly as fast. Of course, the BASIC can handle floating point numbers, which our routine cannot. As you can see, this example not only serves as a good introductory lesson in machine code programming but also illustrates the advantages in compactness and speed you can achieve with machine code routines. You should know that this routine is not quite as fast or as compact as it could have been if you'd used some of the available Z80 commands not yet discussed. Consider it additional incentive to learn more about machine coding. Meanwhile, you're now reasonably equipped to deal with any of the routines you'll find later in this book.

Chapter 5

Tools and Other Unsightly Paraphernalia

The heading of this section may seem somewhat surprising. After all, if you plan to do only projects that don't require hardware, then you won't need any tools. Some of the items below aren't normally considered tools, but you may need them—or at least find them very convenient—in completing some projects.

Needle-nosed pliers. A miniature pair, say the 4-inch size, without side cutters, with plastic insulated grips, and with smooth rather than serrated jaws is best. They're useful as heat sinks while you're soldering, and they're good for holding and placing small wires and parts.

Diagonal cutters. A miniature pair with full flush-cutting jaws and plastic insulated grips are almost essential for neat circuit board work. (Nothing in this book requires more heavy-duty units.)

Wire strippers. You could stick to the old ways and use a knife, diagonal cutters, or scissors, but using strippers will be easier on both the wire and the insulation. A small, continuously adjustable pair like Radio Shack catalog number 64-1952 or 64-2129 will serve you well. But be careful not to adjust the strippers smaller than the gauge of the wire you're using.

Screw drivers. An eighth-inch, flat-bladed screw driver, an eighth-inch Phillips screw driver, and a three-sixteenths-inch flat screw driver, all with plastic grips, should do the job.

Soldering iron. Almost any 25-watt soldering pencil with a fine tip will work. Cleanliness is essential to good soldering, so you might also want a small damp sponge and a bronze brush to clean the tip of your iron.

That completes the list of usual tools. You may have some or all of them already. Substitutions are permissible and additions optional. If you're a novice at the circuit assembly game, you may want to try the tools you have on hand, note what their shortcomings are, and then go to your local tool dealer to see what will make the job easier. Tools are a bit personal, so get what you feel comfortable using. A satisfactory blend of all but the strippers constitutes the Radio Shack "Science Fair" tool set for $14.95. Other dealers probably have similar packages.

A rather special tool is a prototyping board, or "modular IC breadboard sockets," as Radio Shack calls them. The small board is adequate for any project in this book, but if you think you may want to tackle some bigger projects, splurge and get the larger one now. With either, you have the added advantage of being able to transfer your completed circuit to a matching copper-clad board for permanent assembly.

For the sound projects you'll need an amplifier. You can use almost any unit with a 1–3 millivolt sensitivity, which means the magnetic phono input on the family stereo will work, but beware putting a signal into it with the volume turned up: you may blow your speakers out of their boxes. An alternate solution is to use Radio Shack's battery operated amplifier/speaker (catalog no. 277-1008). This device is quite small and portable, and you can place it right next to the computer. If you've already done some electronics, you may want to build the amplifier; numerous schematics and kits exist.

In addition, you'll need a power supply for the external circuit projects, but you'll find the details in a separate section.

You'll also need a few supplies.

Solder. Rosin core, definitely not acid, is best. Buy a 0.032-inch 60/40 alloy.

Wire. For your prototyping board you'll need small spools of #22 solid-core hook-up wire in three or four colors. You can also use it for permanent assembly. If you're going to solder the wire, try to get some with insulation that's reasonably heat-resistant. Radio Shack wire is lacking in this respect, and none of theirs is tinned, a desirable feature.

Electronic components. You can buy these individually for the particular project you wish to build, but if you're planning to build a number of circuits (both from this book and elsewhere), you may be money ahead to obtain some of the pre-packaged assortments and/or "grab bags" of resistors, capacitors, and diodes most electronics parts suppliers carry.

Chapter 6

It's About Time

Without further delays, then, it's down to business. It's about time, you say? This chapter certainly is. And speaking of delays, you'll deal with them, too.

Time is the pacing parameter in all digital computers. It determines the rate at which their internal machinations are carried out and the moment when the computer can or will exchange data with the outside world. You might imagine a computer as a giant array of switches opening and closing according to a sequence imposed by the hardware of the machine and the instructions of the program. An oscillator (or oscillators) controls the rate at which the switches open and close. Commonly referred to as the computer "clock," the oscillator runs as fast as possible, subject to the hardware limitations of the computer circuitry itself. For most current 8-bit machines, this is in the low-to-medium megahertz range.

Each microprocessor has a limited set of operations, or instructions, which it can carry out. Referred to as the machine code instructions, these instructions are available to the particular microprocessor (mP) and are usually expressed as assembly language mnemonics (LD A,n—directly load the accumulator or A-register with the number n). Each instruction requires a specified number of clock cycles, generally in the range of 5–20. Obviously, the faster the clock, the shorter the time required for each instruction. Further, with short machine code routines it is practical to count up the total number of clock cycles required to complete the routine and determine to within the accuracy of the clock exactly how long the routine will take to execute. Because the commands, operations, and functions of high-level languages like BASIC, FORTRAN, COBOL, and FORTH are themselves collections of machine instructions, it is still possible in principle to count up the total number of cycles needed. But the sequence of instructions is so long and includes so many branch points that this approach is impractical. It is practical, however, to time how long, say, 1000 BASIC instructions take. Dividing that time by 1000 then yields an approximate time for a single operation. You can make use of both techniques.

Before you can start building programs you must have a few numbers. Start with the clock itself. You will hear it said or you may have read that the clock in the TS 1000 runs at 3.25 MHz (megahertz). Actually, it runs at about 3.23 MHz. If it ran exactly at 3.23 (or at any other frequency), there would be no problem. However, the master oscillator, which is built around a ceramic filter, wll drift a lot during the first hour after the computer is turned on. During the first hour it can gradually shift upward from 3.230680 MHz to 3.231372 MHz in frequency. Thereafter, the machine will probably settle down somewhat and remain at 3.231200 MHz plus or minus 100 Hz (hertz). If you're programming on your computer during this period, you may find some additional variation. The important points to remember are that if you are using your computer in time-critical situations, let it warm up for at least an hour; and afterwards expect variations of up to 6 parts per 100,000 during one hour. (This drift could amount to 0.2 seconds of error in a tenth-second stopwatch.) For much longer periods, the variations may average out.

The next item you need is some idea of how long each of the command and functions in the TS 1000 BASIC take. Table 6.1 on the following page lists the approximate times for most of them. If one you need is missing, you can still determine it. First, time 1000 iterations of the basic **FOR-NEXT** loop.

```
10 FOR I = 1 TO 1000
20 NEXT I
```

Use a one-hundredth of a second stopwatch. Run it in FAST mode. The time should be very close to 4.46 seconds. To get a better estimate, measure it three or four times, and average your readings. Now go to SLOW mode and repeat the measurement. The time should now be 26.79 seconds. The ratio of the two is 6.007, demonstrating that your computer runs about six times slower in SLOW mode than in FAST, not four times slower. Call these times TLfast and TLslow. Now insert the following line:

```
15 LET X = 1
```

Run this short program in both FAST and SLOW modes. The times should be 5.94 and 36.06 seconds. Call these TXfast and TXslow,

TABLE 6.1

Execution Times for TS 1000
BASIC Commands and Functions
in Fast Mode

Function or Command	Time per Operation *(in milliseconds)
FOR-NEXT (STEP DOESN'T MATTER)	4.46
LET X = 1	1.48
LET X = 1.1	1.60
LET X = 1.12345	1.76
LET X = 11.12345678912	2.03
RAND	1.03
PRINT AT 11,10;1	110.80
PRINT AT 11,10;''A''	3.54
POKE 16514,28	2.00
POKE J,28 (J = 16514)	2.20
LET X = 1 + 0	2.47
LET X = 1 + 123	2.69
LET X = 1 + 1.23	2.64
LET X = 1.23 + 2.34	2.75
LET X = 123 + 234	2.69
LET X = 1 − 0	2.15
LET X = 2.31 − 1.23	2.88
LET X = 1 ✱ 0	2.32
LET X = 1.45 ✱ 1.23	3.89
LET X = 2/1	4.36
LET X = 12.3/2.34	4.53
RAND USR 16514 (16514 CONTAINS 201 = RET)	2.33
LET X = USR 16514 (16514 CONTAINS 201 = RET)	2.84
RAND USR 16514	2.33

```
    CONTENTS:
    LD B,1
    DJNZ,−2
    RET
```

RAND USR 16514	3.35
CHANGE TO LD B,255	
RAND USR 16514	2.37
CONTENTS:	
LD DE,255,255	
LD HL,0,0	
ADD HL,DE	
JR C,−3	
RET	
RAND USR 16514	4.16
CHANGE TO LD HL,255,0	
RAND USR 16514	20.58
CHANGE TO LD HL,0,10	
RAND USR 16514	22.43
CHANGE TO LD HL,0,11	
SCROLL	2.18
CLS	1.57
GOSUB 25	2.10
LINE 25 IS RETURN	
PLOT 11,11	2.36
UNPLOT 11,11	2.36
LET X = SIN .75	40.46
LET X = INT 1.23	2.76
LET X = RND	13.26
LET X = STR$ "2"	112.71
LET X$ = "2"	3.56
LET X$ = CHR$ 29	4.94
LET X = CODE "2"	2.57
LET X = PEEK 16516	2.87
LET X = ARCSIN .6	177.69
LET X = SGN −1	2.45
LET X = ABS −1	2.39
LET X = SQR 3.6	108.65
LET X = VAL "2"	6.70
LET X = LEN "1234"	2.69
LET X$ = INKEY$	3.64
LET X = EXP 3	42.56
LET X = LN 3	67.32

*Each value is an average of three measurements of 1000
operations, with the exceptions of PRINT AT 11,10;1, LET

X\$ = STR\$ ''2'', LET X = ARCSIN .6, LET X = SQR 3.6, and LET
X = LN 3, all of which are single measurements. Extreme spread
for each of the sets of measurements is less than or equal to
0.05 milliseconds.

---

respectively. Note that the ratio of the two is 6.071. The time
required for a single LET X = 1 operation is then

$$(TXfast \text{ or } TXslow - TLfast \text{ or } TLslow)/1000$$

depending on whether you were using the fast or slow times. All
of the values in the table derive from this formula. Each value is
the average of three measurements in FAST mode except as
otherwise noted. The extreme spread for each set of three times
was less than or equal to 0.05 seconds, so the entries are fairly
accurate. You'll find only FAST mode values in the table because
the ratio of Tslow to Tfast is nearly constant for all measurements,
its average value being 6.075. To find Tslow for any operation,
simply multiply the FAST time by 6.075.

Now you're ready to start building a one-second interval, 24-
hour clock/count-up timer program. To understand why the clock
is structured as it is, you'll first conduct three short experiments.
Type in the following program:

```
100 LET X = 0
110 PRINT AT 11,10;X
120 LET X = X + 1
130 GOTO 110
```

RUN this program, and time how long it takes to print the first
100 integers. Add 19 empty REM statements. The program will
now look like this.

```
 5 REM
10 REM
15 REM
20 REM
25 REM
30 REM
35 REM
40 REM
```

```
45 REM
50 REM
55 REM
60 REM
65 REM
70 REM
75 REM
80 REM
85 REM
90 REM
95 REM
100 LET X = 0
110 PRINT AT 11,10;X
120 LET X = X + 1
130 GOTO 110
```

Again, time how long the first 100 integers take. Why has the time increased by a second? After all, the REMs do nothing, and they are not even included in the program loop. The answer is that every time the **GOTO** statement is encountered the computer scans upward in memory from the first line number, starting in location 16509, until it finds the line number given in the **GOTO**. And, of course, you inserted 19 more line numbers between 16509 and the location of line number 110.

Now delete the REM statements, and make the program read:

```
100 LET X = 1
110 PRINT AT 11,10;B$(X)
120 LET X = X + 1
130 GOTO 110
130 GOTO 110
200 DIM B$(10,1)
210 FOR I = 1 TO 10
220 LET B$(I) = STR$(I - 1)
230 NEXT I
```

Begin by entering:

```
FAST
GOTO 200
```

When the computer halts, type:

```
SLOW
GOTO 100
```

See how fast the first 10 integers are printed before the program stops again! If you look at the times for printing a number versus printing a character in Table 6.1, you will get some idea why the integers printed so fast. Your computer handles numbers and strings differently. Numbers are stored in binary form and must be translated to characters for printing. To put it another way, in setting up the string array B$( ) the computer has already done the math and translation, and it does it only once—before the counter is actually started.

Type **NEW** and enter the next program. The directions in parentheses mean do what is inside the parentheses. Don't type the literal contents or the parentheses.

```
10 FOR I = 28 TO 37
20 POKE J,I
30 NEXT I
40 GOTO 10
100 PRINT AT 11,10;"(7 spaces)"
110 LET J = 23 + PEEK 16396 + 256 * PEEK 16397
120 GOTO 10
GOTO 100
```

Why use I = 28 to 37? Because you can't **POKE** characters into memory but 28–37 are the codes for the numerals 0–9.

Why does this routine print even faster? Because **POKE** is a faster operation than **PRINT AT** and the printing that establishes the display file is done in line 100 before the counter begins.

Okay, now on with the clock. Type **NEW** and enter the program below. Be sure to use the line numbers.

```
20 PRINT AT 11,10;B$(H) + ":" + B$(M) + ":" + B$(S)
50 LET S = S + 1
60 IF S = 61 THEN GOSUB 110
100 GOTO 20
110 LET S = 1
120 LET M = M + 1
130 IF M = 61 THEN GOSUB 150
140 RETURN
150 LET M = 1
160 LET H = H + 1
170 IF H = 25 THEN LET H = 1
180 RETURN
460 INPUT H
```

```
 480 INPUT M
 490 LET H = H + 1
 500 LET M = M + 1
 510 LET S = 1
 540 IF INKEY$ < > " " THEN GOTO 20
 550 GOTO 540
1000 FAST
1010 DIM B$(60,2)
1020 FOR I = 1 TO 60
1030 LET B$(I) = STR$(I – 1)
1040 IF I < 11 THEN LET B$(I) = "0" + STR$(I – 1)
1050 NEXT I
1060 SLOW
1070 GOTO 460
```

Lines 20–180 are the basic clock/timer. Lines 460–550 let you zero the timer or set the clock and signal it when to start. Lines 1000–1070 establish the string array of clock characters. Start the program with a **GOTO** 1000, not **RUN**. When the L-cursor appears (after several seconds), enter a zero followed by another zero at the second L-cursor. The screen will now be all white. Hit any key except BREAK to start. Hit BREAK to stop and **CONT**, **ENTER** to continue. *Never use RUN*; it will destroy the string array, and you will have to start over again. Note that your screen is counting too fast. Hit BREAK again, and type the first column of lines listed at the end of this paragraph. Do not type the second and third columns following the **POKE**s. If you wish, you can use the machine code loader given in Chapter 3 instead of individually **POKE**ing each code, but you'll have to delete the code loader later to make room for the rest of the program.

```
10 REM 1234567890
70 FOR K = 1 TO 25
80 NEXT K
90 RAND USR 16514
```

POKE 16514,17	LD DE	Load –1 in the
POKE 16515,255	LOW-ORDER BYTE	DE register pair
POKE 16516,255	HIGH-ORDER BYTE	low-order byte first.
POKE 16517,33	LD HL	Load 1 in the
POKE 16518,1	LOW-ORDER BYTE	HL register pair
POKE 16519,0	HIGH-ORDER BYTE	low-order byte first.

```
POKE 16520,25      ADD HL,DE              Add DE to HL,
                                          store result in HL.
POKE 16521,56      JR C                   Jump back a step if
                                          carry resulted;
                                          otherwise,
POKE 16522,–3      JUMP SIZE,BACK 3       continue.
POKE 16523,201     RET                    Return to BASIC.
GOTO 20
```

Notice how much slower your program is counting. Why? You put in two delays. The first is in lines 70 and 80, where the computer just loops 25 times in BASIC doing nothing. The second is the machine code routine you **POKE**d into the REM in line 10.

The assembly language mnemonics for the machine code itself appear in the column to the right of the **POKE** statements, while the third column contains a description of what the code does. Briefly, in English, it adds –1 to the number in HL, it stores the result back in HL, it notes whether or not a carry bit was generated, and it repeats this process as long as the carry bit is generated. In your case, this was only once, since you loaded the number one into HL. If there is no carry, then the machine code program carries straight through the JR C. See your computer's instruction manual for additional information. If you wish to increase your machine coding skills—definitely worthwhile if you want to get the most from your machine—buy one of the many books available on machine language programming for the ZX81 or Z80/Z80A microprocessor.

Now, back to the clock. Forget about timing for the moment. You no longer need lines 1000–1070, since the string array is already generated, so delete them. At this point you should also **SAVE** your program so that if you accidentally start with **RUN** you won't have to regenerate it. Those of you with 1K memories now have room for a visual alarm in your clock: those with 2K have no problems. Add these lines.

```
30 IF H=X AND M=Y AND S=Z THEN PRINT AT
   13,9;"TIME IS UP"
360 INPUT X
380 INPUT Y
400 INPUT Z
410 LET X=X+1
420 LET Y=Y+1
430 LET Z=Z+1
```

```
GOTO 360
```

The L-cursor will come up five times. Enter alarm hour, alarm minute, alarm second, set hour, and set minute in that order. The screen will go white waiting for you to push any key to start the clock/timer. When the alarm time is reached, ''TIME IS UP'' will print under the count, and the clock will continue counting. If you don't want the alarm to go off, then enter a number >25 at the first L-cursor. To reset the alarm you must restart the clock.

If you have only 1K, you can go no further (though you may want to delete the alarm and use some of the features further below), so you can work on timing the clock. Time how long it takes your clock to count to one minute. If it is fast, increase the 25 in line 70; if it is slow, decrease it. From Table 6.2 you can see that each increment of the upper limit in line 70 should change the timing by about 27.1 msec (millisecond) per second (4.46 msec for fast times 6.075 to get to the slow time), or 1.6 seconds per minute. Find the value that lets the clock count to 00:01:00 in just under but not over one minute. Now **POKE** 16519 with 0, 1, 2, . . . and so on until you find a value that gets as close to but less than one minute as possible. If you have difficulty measuring the time, increase the period to 3 or 5 minutes. Table 6.1 tells you that the time should change about 0.67 second per minute as 16519 is incremented. Now you're at the last timing element. Begin **POKE**ing 16518 with 1, 2, . . . until the clock keeps as accurate time as possible. You will have to measure the time over an hour or more now, and eventually overnight, since each increment of 16518 changes the timing by only 0.15 second per hour. When properly set, the clock should keep time to about 5 seconds per day. Theoretically, by adding NOPs to your machine code timing loop you could get it to keep time to 0.65 seconds (1.24✻6.075 microseconds per second) per day in SLOW mode, but without being sure of the 3.23 MHz clock stability this seems a futile exercise. If you choose, you can also delete lines 70 and 80, and time your clock entirely with the machine code routine. The **FOR-NEXT** loop was added only to show how you can use do-nothing loops, whether machine code or BASIC, for timing and how you can use different types of loops for coarse and fine tuning.

If you have 2K or more memory you can add some convenience touches to your clock. Add these lines.

```
40 IF INKEY$ < > '' '' THEN PRINT AT 13,9;''(10 spaces)''
300 LET X = 0
```

```
310 PRINT AT 9,7;"WANT ALARM? Y/N"
320 INPUT A$
330 IF A$ < > "Y" THEN GOTO 440
340 PRINT AT 9,7;"SET ALARM (5 spaces)"
350 PRINT AT 11,7;"ENTER HOUR"
370 PRINT AT 11,13;"MINUTE"
390 PRINT AT 11,13;"SECOND"
440 PRINT AT 9,7;"SET TIME (6 spaces)"
450 PRINT AT 11,7;"ENTER HOUR(2 spaces)"
470 PRINT AT 11,13;"MINUTE"
520 CLS
530 PRINT AT 11,10;"START ME"
GOTO 300
```

If you added line 40, you'll have to retime your clock, since it's within the counter loop. The author ran the complete clock with lines 70 and 80 deleted and found that it gained 18 seconds in 13 hours, with 61 in 16518 and 66 in 16519. The complete clock program is listed as Program 6.1.

As an exercise, you can convert the clock to an a.m./p.m. clock or a countdown timer. You can do the conversions in BASIC, and you can use the same machine code delay, although probably with different timing parameters. Hint: the basic timer will count down if you decrement rather than increment H, M, and S, and change the tests in lines 60, 130, and 170. Presumably, you will fix the alarm at zero.

PROGRAM 6.1 24-Hour Clock/Timer

```
10 REM 1234567890
PRINT AT 11,10;B$(H) + ":" + B$(M) + ":" + B$(S)
30 IF H = X AND M = Y AND S = Z THEN PRINT AT
   13,9;"TIME IS UP"
40 IF INKEY$ < > " " THEN PRINT AT 13,9;"(10 spaces)"
50 LET S = S + 1
60 IF S = 61 THEN GOSUB 110
90 RAND USR 16514
100 GOTO 20
110 LET S = 1
120 LET M = M + 1
130 IF M = 61 THEN GOSUB 150
140 RETURN
```

```
150 LET M = 1
160 LET H = 1 + 1
170 IF H = 25 THEN LET H = 1
180 RETURN
300 LET X = 0
310 PRINT AT 9,7;"WANT ALARM? Y/N"
320 INPUT A$
330 IF A$ < > "Y" THEN GOTO 440
340 PRINT AT 9,7;"SET ALARM(5 spaces)"
350 PRINT AT 11,7;"ENTER HOUR"
360 INPUT X
370 PRINT AT 11,13;"MINUTE"
380 INPUT Y
390 PRINT AT 11,13;"SECOND"
400 INPUT Z
410 LET X = X + 1
420 LET Y = Y + 1
430 LET Z = Z + 1
440 PRINT AT 9,7;"SET TIME(6 spaces)"
450 PRINT AT 11,7;"ENTER HOUR(2 spaces)"
460 INPUT H
470 PRINT AT 11,13;"MINUTE"
480 INPUT M
490 LET H = H + 1
500 LET M = M + 1
510 LET S = 1
520 CLS
530 PRINT AT 11,10;"START ME"
540 IF INKEY$ < > " " THEN GOTO 20
550 GOTO 540
1000 FAST
1010 DIM B$(60,2)
1020 FOR I = 1 TO 60
1030 LET B$(I) = STR$(I − 1)
1040 IF I < 11 THEN LET B$(I) = "0" + STR$(I − 1)
1050 NEXT I
1060 SLOW
1070 GOTO 300
```

Sequentially **POKE** in the following codes, beginning at location 16514:

17, 255, 255, 33, 61, 66, 25, 56, 253, 201

Start with GOTO 1000

You've learned quite a bit in assembling your clock, but there's another very important timing loop you should be familiar with, and to find out about it you'll build a tenth-second stopwatch. Type in Program 6.2. You can directly **POKE** in the machine code routine as shown below. Begin the program with a GOTO 200. After "ST:AR.T" appears, press any key except BREAK to start the timer. Use BREAK to stop it.

PROGRAM 6.2 Tenth-Second Stopwatch

```
 10 REM 123456789
 20 FOR A = 28 TO 37
 30 POKE 16894,A
 40 FOR B = 28 TO 37
 50 POKE 16895,B
 60 FOR C = 28 TO 33
 70 FOR D = 28 TO 37
 80 FOR E = 27 TO 37
 90 POKE 16897,C
100 POKE 16898,D
110 POKE 16900,E
120 RAND USR 16514
130 NEXT E
140 NEXT D
150 NEXT C
160 NEXT B
170 NEXT A
200 PRINT AT 11,10; "ST:AR.T"
210 IF INKEY$ < > " " THEN GOTO 20
220 GOTO 210
```

POKE 16514,6	LD B	Load B register with N.
POKE 16515,14	N	Value to load B with.
POKE 16516,16	DJNZ	Decrement B, jump back 2 if not zero;
POKE 16517,–2	JUMP SIZE	otherwise, pass through.
POKE 16518,201	RETURN	Return to BASIC.

Sequentially **POKE** in the following codes beginning at location 16574.

6, 14, 16, –2, 201

Start with GOTO 200

Add 358 to the **POKE** locations in lines 30, 50, 90, 100 and 110 if 16K memory is used.

In order to make a tenth-second stopwatch run in BASIC, a concession to accuracy had to be made. If you watch the tenths scroll by carefully, you'll note that the digit 9 takes slightly longer than any of the others. This is because at least two and sometimes as many as five **FOR-NEXT** loops all increment as the 9 flips to a 0. The stopwatch has a maximum error (not counting clock instabilities) of 0.091 second in any given second, but the average error is only 0.045 second. The author timed it several times with a hundredth-second stopwatch and observed a total error of 0.1 second over a 90-minute period.

The delay used in this program is a very simple but very powerful routine. It begins when you load the B register with 0 or any integer from 1 to 255. You **POKE**d in 14; 0 effectively makes B act as though loaded with 256. The next instruction, DJNZ (decrement, jump if not zero), decrements B, compares the result with 0 and jumps the number of bytes following DJNZ if the result was not 0. If it was 0, program execution drops through to the next instruction. Note that the decrement takes place before the comparison. This is why 0 acts like 256; decrement 0 and you get –1, which, in binary, is 8 ones or 255. In your program, the jump is two backwards, so the program counter in the microprocessor is right back at the DJNZ instruction. In other words, the computer just wastes time by decrementing B until it reaches 0, and then it proceeds on its way.

The stopwatch is fine-tuned much as the clock was. Vary the value of 14 in location 16515 until the watch just barely gains time in whatever interval you choose. Each increment of 16515 will change the timing about 0.85 second per hour. Because you can bracket the correct delay, the actual error can never be more than half this, however. Still, it's too great an error for a 99-minute stopwatch. You have four extra characters in your REM statement, so what you will do is **POKE** in up to three NOP instructions. An NOP tells the microprocessor to do nothing once, and it takes 1.238 microseconds to run. This translates to 0.271 seconds per hour in our program (1.238 seconds in fast times 6.075 seconds for slow times 36000 for the number of times it occurs in one hour.) So now try

POKE 16518,0
POKE 16519,201

Always remember to **POKE** 201 into the next higher location so that the machine will return to the BASIC program. Is your watch keeping better time? If it's still too fast, try adding another NOP in location 16519. Remember to **RET**urn. If you have to add more than three NOPs then you should remove them all, increment 16515, and try again.

Delays are so important in so many programs that you should acquaint yourself with the three different delays you'll find in Table 6.2. The table also lists the number of clock cycles the delays take, and it tells you how to compute the actual time they take. The numbers adjacent to the mnemonics in the table are the decimal codes that you should **POKE** into the computer to realize the delays.

C is the number of clock cycles. D is the maximum delay time, and T is the increase in delay for each increment of the counter. The time in fast mode for any delay is

$$\text{Time} = \frac{C * .000001}{3.23}$$

You've already seen delays B and HL in the stopwatch and timer programs. Delay BC demonstrates the technique of "nesting," wherein one delay is contained within another. The C register is loaded with a number between 0 and 255. The B delay is then carried out. At the end of the B delay C is decremented and examined to see if it is 0. If it's not, then the B delay is again repeated, C is again decremented and tested for 0, and so on until C finally becomes 0. You can see that the total time will be approximately N1 (the number originally loaded in C) multiplied by the duration of the B delay. This works very much like a backwards odometer. The units marker on your car's odometer is incremented ten times for each time the tens marker is incremented once. So the B register is decremented N2 − 1 times for each time the C register is decremented once. The nesting can be as deep as your registers will accommodate; you could, for instance, contain the whole BC delay between the LD HL, HILO and the ADD HL,DE of the HL delay and achieve a total delay of about HILO*N1*duration of B delay. You can also stack or concatenate delays (use them one after the other), in which case the total delay will be the sum of the individual delays.

That concludes the chapter on clocks, timers, and delays. It's about time, isn't it?

TABLE 6.2 Three Useful Delay Routines

MNEMONIC	CODE	
LD B,N	6,N	
DJNZ,−2	16,−2	
$C = 13*N + 2$	$T = .00425$ msec	$D = 1.0246$ msec

DELAY B—simplest, shortest

LD C,N1	14,N1	
LD B,N2	6,N2	
DJNZ,−2	16,−2	
DEC C	13	
JR NZ,−7	32,−7	
$C = 2 + 16*N1 + 13*N1*N2$	$T = .00425*N1$ for N2	
$D = 263.4$ msec	$T = .00495 + .00425*N2$ for N1	

DELAY BC—"nested" delay

LD DE,65535	17,255,255	
LD HL,HILO	33,LO,HI	
ADD HL,DE	25	
JR C,−3	56,−3	
$C = 23*HILO + 28$	$T = .0712$ msec for LO	
$D = 466.7$ msec	$T = 18.23$ msec for HI	

NOTE: If HILO is a number between 0 and 65535, then
       $HI = INT(HILO/256)$ and $LO = HILO - 256*HI$

DELAY HL—16-bit delay

NOTE N, N1, N2 are 8-bit; HILO is 16-bit.

Chapter 7

Sound Advice

Lots of computer users like you want to know whether they can make noise with their machines—just some kind of sound. The answer? No—and yes. No, if you're asking whether there's an accessible BASIC sound command or generator. But, if you reflect for a moment on the raucous sounds that come from your tape recorder when you play back a **SAVE**d tape, you'll realize that your computer is capable of producing some sounds. The staccato buzz you hear preceding the actual **SAVE**d program is the vertical synch pulses that synchronize the TV set and computer. The sound of the **SAVE**d program is a series of short 3.3-KHz tone bursts, four cycles of the tone if a bit is zero and nine cycles if it is one.

In order to hear sounds in real time (without having to record and play back), you'll need an amplifier with a 1–3 millivolt sensitivity. The magnetic phono input on your stereo will do, but when you hook up to it, make sure the volume control is turned all the way down. Better yet, use the Radio Shack amplifier mentioned in Chapter 5. Turn the computer on in SLOW mode and increase the volume until you hear that staccato buzz at the desired level. Type in

```
 5 FAST
10 IF INKEY$ < > " " THEN GOTO 10
20 IF INKEY$ = " " THEN GOTO 20
```

and **RUN** it. The screen should go blank and the buzz should stop. If it does not, then you have grounding problems. Check your cable from the computer microphone (labeled MIC) socket to the amplifier input socket. Is it fully inserted? Did you have to make up a cable and, if so, did you swap the shield and center conductor? If your recorder is connected, disconnect it. Try reversing the line plugs on the TV and amplifier one at a time. If the buzz is still there, you better resign yourself to obtaining a battery-powered amplifier.

You *could* make use of these sounds. Suppose you were running a program in FAST to monitor your heating thermostat. When the computer detected that the device had tripped, you could have it switch to SLOW or **SAVE** the program to produce an audible alarm through your amplifier.

In general, however, you'll want to call up sounds that are more pleasing to the ear. You'll discover how to do that, but you'll always be subject to the restriction that when in SLOW mode the vertical synch pulse buzz will always be present.

The easiest way of making a sound other than the buzz is to use the **SAVE** routine built into your ROM (*Read-Only Memory*). Instead of letting it output signals related to the program you'll load it with a single character and have it repeatedly output. This is one of the shorter programs you're likely to run across. Type in Program 7.1 and **RUN** it in FAST mode. The mnemonics follow.

Starting Location	Codes	Mnemonic
16514	30,255	LD E, 255
16516	205,31,3	CALL 799
16519	24,–7	JR –7
16521	201	RET

PROGRAM 7.1 SAVE-Like Signal

```
10 REM 12345678
20 RAND USR 16514
```

Sequentially **POKE** in the following codes beginning at location 16514

30, 255, 205, 31, 3, 24, –7, 201

This notation, unlike the notation in the last chapter, lists the entire command along with its arguments, if any, on a single line. You should place the first element of the command in the starting location. In the first line, 30 goes in 16514, and 255 goes in 16515. It is a more compact way of writing the routine and is the format most assembler programs for the TS 1000 use.

What it does is load a bit pattern into the E register and then call the tone burst generator portion of the ROM's **SAVE** subroutine. The generator produces four cycles of the 3.3-KHz tone for each zero in the pattern and nine cycles for each one. After the appropriate tone burst has sounded for each bit in the E register, the process repeats. You can change the nature of the sound a little bit by **POKE**ing different numbers between 0 and 255 into the second location, 16515. BREAK will terminate the program. Although not a pretty sound, this signal can be useful in adjusting the volume and tone settings on your tape recorder in preparation for **SAVE**ing a program; it's more efficient than writing a long program and repeatedly **SAVE**ing it until you get the settings right.

Just exactly how does the computer make these noises? It does so by making use of the TV vertical synch circuitry. Every time the microprocessor issues this circuit an output command, the synch signal goes high, and whenever it issues an input command it goes low. The synch signal is heavily filtered and attenuated before it reaches the microphone output of the computer so that what appears there is a low-level signal approximating a sine wave. This filtering is necessary, in part, to eliminate the 5-microsecond horizontal synch pulses generated automatically whenever the vertical synch is high. The Z80 addresses the output command to port 255 and the input command to port 254. These port numbers refer to the 256 ports that the Z80 is capable of sending signals to. Apparently, these port assignments resulted from hardware and cost considerations.

With this information, you can proceed to program a tone generator. Type in Program 7.2. Here are the mnemonics.

Starting Location	Codes	Mnemonic
16514	62,127	LD A,127
16516	219,254	IN A,254
16518	31	RRA
16519	208	RET NC
16520	0	NOP
16521	6,122	LD B,122
16523	16,−2	DJNZ −2
16525	211,255	OUT 255,A
16527	6,122	LD B,122
16529	16,−2	DJNZ −2
16531	24,−19	JR,−19

PROGRAM 7.2 Continuous Tone Generator

```
10 REM (19 characters)
20 FAST
30 RAND USR 16514
```

Sequentially **POKE** in the following codes beginning at location 16514

62, 127, 219, 254, 31, 208, 0, 6, 122, 16, –2, 211, 255, 6, 122, 16, –2, 24, –19

In 16514–16519 the synch is reset, the BREAK key is checked, and a return to BASIC is made if it's depressed. The 16520–16526 delays for 0.5 msec (millisecond) and then sets the synch. The NOP is used to make the on and off portions of the signal symmetric. Locations 16527–16532 delay for 0.5 msec and then jump back to the beginning of the routine where synch is again reset.

The tone is 1-KHz. To change it, **POKE** numbers other than 122 into 16522 and 16528. The frequency will be

$$\text{freq} = \frac{3230000}{26 * N + 63} \text{ where N is the number}$$

The filter mentioned earlier won't let you generate frequencies much below 500 Hz or above 9 KHz, but this should cover most frequencies of interest to you. To produce the tone, simply **RUN** the program; BREAK will halt it.

One way you could use this routine is as a ''fiddle'' alarm. Suppose your computer is quietly sitting there in FAST mode collecting data or performing some other useful task (or meaningless one, for that matter) in BASIC. Include this line in your BASIC program:

XXXX IF INKEY\$ < > '' '' THEN RAND USR 16514

and an audible alarm will be sounded if some one ''fiddles'' with the keys. You can also use this program as an audio test signal generator for many electronics experiments.

Being able to generate tones is one thing, but you'd also like to be able to control them. You'd have a dandy Morse code practice oscillator, for instance, if you could turn the tone on and off from the keyboard. With a few changes, you can. Type Program 7.3. The mnemonics follow.

Starting Location	Codes	Mnemonic
16514	62,0	LD A,0
16516	219,254	IN A,254
16518	31	RRA
16519	208	RET NC
16520	230,15	AND 15
16523	254,15	CP 15
16525	40,–12	JR Z,–12
16527	6,121	LD B,121
16529	16,–2	DJNZ,–2
16531	211,255	OUT 255,A
16533	0	NOP
16534	0	NOP
16535	0	NOP
16536	6,121	LD B,121
16538	16,–2	DJNZ,–2
16540	24,–27	JR,–27

PROGRAM 7.3 Morse Code Oscillator

```
10 REM (27 characters)
20 FAST
30 RAND USR 16514
40 SLOW
```

Sequentially **POKE** in the following codes beginning at location 16514

62, 0, 219, 254, 31, 208, 230, 15, 254, 15, 40, –12, 6, 121, 16, –2, 211, 255, 0, 0, 0, 6, 121, 16, –2, 24, –27

This program is almost the same as the last one. The primary difference is that it checks all of the keys to see if you pressed any. It first checks BREAK in 16518–16519 and then checks the remaining keys in 16520–16523 and jumps back to the beginning to look again if you pressed none. That is, your computer will produce no tone unless a key is depressed, and it will produce the tone cycle by cycle, checking during each cycle to see that a key is still pressed. The additional NOPs and the reduction of the contents of the B register to 121 simply maintain a symmetric 1-KHz tone by compensating for the time the additional key checking takes. The program tests the keyboard in such a way that the keys on the left and right edges of the keyboard will yield a return to BASIC except for BREAK, which will break. Changing the numbers in 16527 and 16536 from 121 will change the frequency. The frequency will be

$$\text{freq} = \frac{3230000}{26 * N + 87} \text{ where again N is the number}$$

You needn't use a keyboard key. If you have a Morse key you can use it by paralleling any of the inner keyboard keys. Chapter 8 has further details.

You'd also like to be able to summon a tone burst from your programs for various purposes and then have the program return to whatever it was doing. You'll build a demonstration program to do this and use it in the example that follows. You can then use it in any program you wish. Type in Program 7.4. Mnemonics follow.

Starting Location	Codes	Mnemonic
16514	17,255,255	LD DE,−1
16517	33,0,1	LD HL,HILO
16520	211,255	OUT 255,A
16522	6,124	LD B,124
16524	16,−2	DJNZ,−2
16526	219,254	IN A,254
16528	6,122	LD B,122
16530	16,−2	DJNZ,−2
16532	0	NOP
16533	25	ADD HL,DE
16534	56,−16	JR C,−16
16536	201	RET

This routine bears marked similarities to the previous two. The basic tone generation routine (16520–16532) is essentially the same, but it omits the machine code checks of the keyboard. That function is now performed in BASIC so that you can use the routine with games or as a warning device once it meets some criterion you've specified. More importantly, the generator is placed within a timing loop. The HL register pair is loaded with a number, HILO, and −1 is added to it during each cycle of the tone signal, decreasing the value in HL by one. So long as HL is nonzero,

a carry will be generated, resulting in a jump backward (16534).
When HL reaches zero, the carry is no longer generated, and
control passes first to RET and from there back to BASIC.

This program will sound a fixed length, fixed frequency tone
whenever you press a key. When you run it, it will ask you for a
frequency number. Given the frequency in hertz, the number is

$$N = (3230000/freq - 79)/26$$

The program will also ask you for a length number, which determines how long the tone burst will be. Given the length, L, in
seconds, HILO will be

$$HILO = \frac{L * 3230000 + 5 - (26 * N + 79)}{26 * N + 79}$$

For HILO = 256 and N = 124 (the values you entered), you will get
about 0.25 second of 1-KHz sound. You can, as an exercise, modify the program so that it asks you for the frequency and duration;
you can then have it compute the values of N and HILO.

Having gone this far, try turning this routine into a metronome.
EDIT line 10 of the previous program so that it contains 17 more
characters following the TAN, and modify your loader so that line
8010 reads

    8010 FOR I = 16536 TO 32000

POKE in these additional codes using the modified loader

    6,19,33,176,12,62,127,219,254,31,208,25,56,–9,
    16,–14,24,–37

In immediate mode

    POKE 16518,49
    POKE 16519,0

The mnemonics that follow are a continuation of those in Program
7.4. (The 6 replaces the 201 in location 16536, and the rest of the
new code takes the place of the additional 17 characters you
added to the REM.

Starting Location	Codes	Mnemonic
16536	6,19	LD B,R
16538	33,176,12	LD HL,3248

16541	62,127	LD A,127
16543	219,254	IN A,254
16545	31	RRA
16546	208	RET C
16547	25	ADD HL,DE
16548	56,–9	JR C,–9
16550	16,–14	DJNZ,–14
16552	24,–37	JR,–37

PROGRAM 7.4 Tone Burst Generator

```
10 REM (23 characters)
20 PRINT "INPUT FREQ NUMBER"
30 INPUT N
40 PRINT "INPUT LENGTH NUMBER"
50 INPUT HILO
60 POKE 16523,N+2
70 POKE 16529,N
80 POKE 16519,INT (HILO/256)
90 POKE 16518,HILO – 256 * INT (HILO/256)
100 IF INKEY$ < > " " THEN RAND USR 16514
110 GOTO 100
```

Sequentially **POKE** in the following codes beginning at location 16514

17, 255, 255, 33, 0, 1, 211, 255, 6, 124, 16, –2, 219, 254, 6, 122, 16, –2, 0, 25, 56, –16, 201

What you've done is to change the length of the tone burst to 0.05 second (16518,16519) and add an R*.05-second-long delay (16536–16551) during which the computer makes no sound. The routine then jumps back (16552,16553) and repeats. What you get is a tone burst 0.05 second long every .05*(R+1) second. You'll get the tone, and the TV screen will flicker in time with it, providing you with a visual as well as an audible metronome. Delete lines 20-110 of the previous BASIC program (*do not delete the REM*), and enter lines 20–100 of Program 7.5.

By inputting values from 0.1 to 12.8 you can vary the beat from 10/sec (10 per second) to 0.08/sec. Realizing that sec/beat is the reciprocal of beats/sec you can change program line 30 to ask for beats/sec if you like and then let R equal its reciprocal in a new line, 45. Run the program in FAST mode and use BREAK to halt it.

PROGRAM 7.5 Metronome

```
 10 REM (40 characters)
 20 PRINT AT 11,7;"INPUT SEC/BEAT"
 30 INPUT R
 40 IF R<.1 OR R>12.8 THEN GOTO 90
 50 POKE 16537,INT (−.5+R/.05)
 60 FAST
 70 RAND USR 16514
 80 STOP
 90 PRINT "VALUE OUT OF RANGE. TRY AGAIN"
100 GOTO 20
```

Sequentially **POKE** in the following codes beginning at location 16514

17, 255, 255, 33, 49, 0, 211, 255, 6, 124, 16, −2, 219, 254, 6, 122, 16, −2, 0, 25, 56, −16, 6, 19, 33, 176, 12, 62, 127, 219, 254, 31, 208, 25, 56, −9, 16, −14, 24, −37

Metronomes are very useful devices for checking the rhythm of many relatively slow processes—sort of a low-speed stroboscope. Probably the area where they are most useful, though, is music. Can your computer whistle a tune? Well, it may not win the "Instrumental of the Year" award, but it will crank out a recognizable melody. You'll use the machine code you already have and modify it by

```
POKE 16536,201
POKE 16518,0
POKE 16519,1
```

Notice that what you've done is to put a RET (to BASIC) in location 16536 and a value of 256 for HILO into locations 16518 and 16519, causing the tone length to be a nomiral 0.25 sec. This converts the tone length back into the code you used in the program before last, but it leaves the rest of the code intact so you can convert back to the metronome by re-**POKE**ing the three locations you just entered. Now delete everything but the REM statement and type in lines 20–70 of Program 7.6. Run the program in FAST mode and use BREAK to halt it.

With this program, the 0 key is an A at 880 Hz. The keys then correspond to the notes as below.

Key	0	1	2	3	4	5	6	7	8	9	A	B	C	D	E
Note	A	A#	B	C	C#	D	D#	E	F	F#	G	G#	A	A#	B

If you substitute 201.87 for the 142.04545 in the program, then the keys all shift down 6 notes so that the 0 key is the D# below the A at 880 Hz. The notes are all fixed-length, and the lengths decrease slightly with increasing frequency. If you can put up with this and with a keyboard that certainly was not meant for playing music, then you can pound out a tune.

Can you think of a way of combining this BASIC program with the machine code from Program 7.3 so that a note would sound for as long as you held down the key (Hints: notice that signal timing is controlled by locations 16527 and 16536 in Program 7.3, not by 16523 and 16529. Recall that the timing numbers are the same, not different by 2. The Program 7.3 machine code does not return to BASIC unless you press one of the edge keys. After you press the first key, the code will always play the same note unless it can get back to BASIC to find out what the value of Y is. It does, however, check to see if you pressed any key. Suppose that instead of jumping back 12 to look again when it found no key, the program were to jump ahead to location 16541 into which you had cleverly placed a RET?)

PROGRAM 7.6 Keyboard Music

```
10 REM (23 characters)
20 IF INKEY$ = " " THEN GOTO 20
30 LET Y = INT (-2.5 + 142.04545 * 1.0596431 * * - (CODE
   INKEY$ - 28))
40 POKE 16523,Y + 2
50 POKE 16529,Y
60 RAND USR 16514
70 GOTO 20
```

Sequentially **POKE** in the following codes beginning at location 16514

17, 255, 255, 33, 0, 1, 211, 255, 6, 124, 16, –2, 219, 254, 6, 122, 16, –2, 0, 25, 56, –16, 201

Do you have perfect pitch? What happens if the value of 2.5 is changed in the expression for Y?

Perhaps a better approach is to forget about the keyboard and write your music into the computer. Change lines 20, 30, and 70 to the statements shown below.

```
20 FOR I = I TO 25
30 LET Y = INT (−2.5 + 142.04545 * 1.0596431 * * −
   (CODE A$(I) − 28))
70 NEXT I
```

Then in the immediate mode type in a string called A$ that is 25 characters long.

```
LET A$ = "1122345...779"
```

Make each character correspond to the note you want the computer to play according to the scale given above. Hit **ENTER** and then **GOTO** 20. If you start the program with **RUN** you will destroy A$. If the tempo is too slow or fast for you, **POKE** 16519 and 16518 with different values. The tone length will increase about 0.25 second for each increment of 16519 and about 0.05 second for a change of 50 in 16518. If your tune is longer or shorter than 25 notes, change the upper limit in line 20 to match the number. If you want the song to play continuously then add 80 **GOTO** 20.

You're still not satisfied because you can't change the length of the notes? Okay, then, you died-in-the-wool music lover, here's one more change.

```
20 FOR I = I TO 50 STEP 2
55 POKE 16519,CODE A$(I + 1) − 46
```

The complete program with machine code and line 80 appears in Program 7.7.

Now input your song as:

```
LET A$ = "1J2J3J2K...7L9L"
```

Here the first character is a note, the second the length of that note, the third a note, the fourth its length, and so on. J is the shortest note, K is twice as long, L is three times as long, and so on. You must include both the note and the length of each. If your tune isn't exactly 25 notes long, then make the upper limit in line 20 equal to twice the number of notes. Remember, this change can be made even after you've typed in A$. You can type as many songs as you'd like into A$ as long as you make the limit in 20 equal to twice the total number of notes.

This doesn't solve all the problems with making music (the notes still get shorter with pitch, there is no provision for rests, and the notes aren't quite on pitch, to name a few), but you need something to do on cold winter nights, don't you?

PROGRAM 7.7 Music from Memory (Player Piano)

```
10 REM (23 characters)
20 FOR I = 1 TO 50 STEP 2
30 LET Y = INT (-2.5 + 142.04545 * 1.0596431 * * - (CODE
   A$(I) - 28))
40 POKE 16523, Y + 2
50 POKE 16529, Y
60 POKE 16519, CODE A$(I + 1) - 46
70 RAND USR 16514
80 NEXT I
```

Sequentially **POKE** in the following codes beginning at location 16514

17, 255, 255, 33, 0, 1, 211, 255, 6, 124, 16, -2, 219, 254, 6, 122, 16, -2, 0, 25, 56, -16, 201

Chapter 8

At the Touch of a Key

The projects in this chapter require internal additions or modifications to your computer. Although there is little danger that the computer will be damaged if you follow these instructions and use a little common sense, it is almost certain that opening the TS 1000 will void any guarantee that covers it. If you are new to electronics construction, you may want to gain some experience with the projects which require only construction external to the computer. In either case, you may wish to postpone working through this chapter until a later date—until the guarantee has expired, anyway, or until you have some more experience. When you *are* ready, the first thing you'll have to do is open the computer. You can accomplish this part easily.

Place the computer upside down in front of you with the printing on the back oriented so that you can read it. With a knife or small screwdriver, pry off the lower two and the upper left rubber foot pads. With an 1/8-inch Phillips screwdriver, remove the five screws in the bottom of the case. You can then lift off the bottom and expose the underside of the printed circuit (PC) board.

To remove the PC board, unscrew the remaining two Phillips screws. The PC board is now free, the keyboard still attached. If possible, do whatever modifications you plan without disconnecting the keyboard, but be very careful not to tear the plastic connecting leads. If you must disconnect the keyboard, lift and rotate the PC board toward you and carefully pull the two plastic leads from their sockets at the lower left of the PC board. (*Don't* forget to reconnect later when you reassemble.) If you should tear either or both leads near the connectors, there's enough slack so that you can cut off the torn part with a pair of scissors and carefully reshape the remainder so you can plug them back in.

Note: Reassembly is just the reverse of the foregoing steps. (If the rubber feet won't stay attached to the case bottom, try a little contact cement.)

Now that the case is open, take a look at that keyboard. Whether your goal is making program entry easier, making games more convenient, or simply enjoying the tinkering, you'll find the additions of a push-button keyboard and joystick worthwhile.

The ZX81/TS 1000 keyboard is a scanned array of 40 switches. You can easily substitute any other switch or array of switches for each or all of the existing keys. Your computer will consider a joystick (of the Atari type, not potentiometric varieties requiring multichannel A/D (*Analog-Digital*) converters for input) nothing more than a partial array.

To see how this works, look at the keyboard diagram, Figure 8.1. Notice that the four rows of the keyboard are split down the

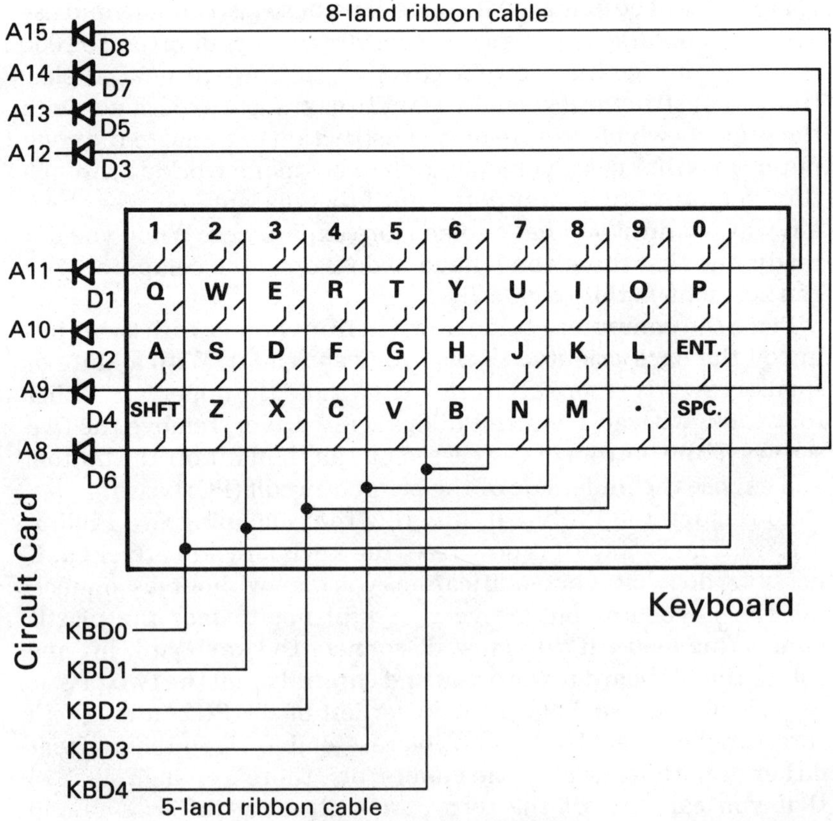

FIGURE 8.1 Keyboard Wiring

middle to form eight half-rows of five keys each. These half-rows are fed by the eight address lines A8–A15 through the diodes D1–D8 (not in the same numerical orders).

Each of the keys (actually, the ''other side'' of each switch, which is the key) in each of the half-rows is also attached to one of the KBD0–KBD4 lines. These lines are normally held at +5 volts through a ''pull-up'' 10K-ohm resistor. To see what happens next, look at a particular switch, say W, which represents the intersection of the A10 and KBD1 lines.

With the switch open, KBD1 will remain at +5 volts regardless of the voltage on A10. However, if the switch is closed and you put 0 volts on A10, then D2 will conduct, and KBD1 will drop to about 1 volt. Thus, by sequentially placing 0 volts on the address lines A8–A15 (only one at a time) while simultaneously examining KBD0–KBD4, you can tell which, if any, of the keys are depressed. In fact, this is exactly what the ROM in your computer causes the microprocessor to do. (It also ''de-bounces'' the keys and checks for multiple closures, but that's another story.)

It's easy to see from Figure 8.2 that you could remove W and substitute another switch by running leads from KBD1 and D2 or even just parallel another switch with W.

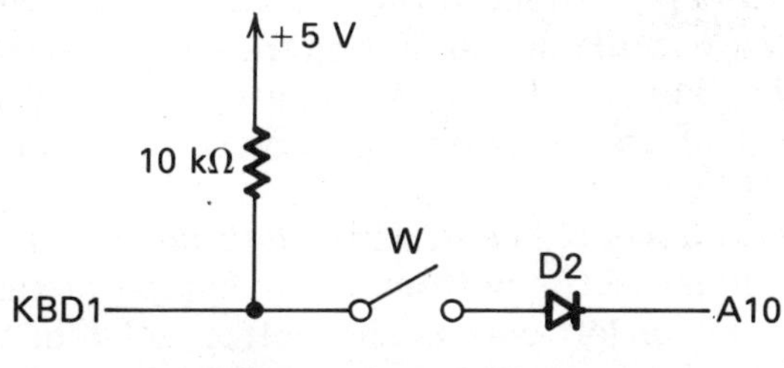

FIGURE 8.2 Switch W

To provide an external, full-size keyboard, then, you need only provide yourself with a collection of 40 switches and wire them according to Figure 8.2. That is, line A11 to one side of switches 1–5, the other side of each switch to lines KBD0 through KBD4 in order, line A10 to one side of switches Q, W, E, R, T, the other side of each switch to lines KBD0 through KBD4 in order, and so on through the entire array. You do not have to leave the switches in the same physical arrangement as the original, although this may be most convenient. For instance, you might want to group the first row (1–0) as a separate numeric keypad, bearing in mind the additional functions of these keys. Or you might want to provide 10 additional keys as a separate keypad. You can do this by paralleling the 10 extras on a one-for-one basis with keys 1–0 of the first row.

Although you could purchase individual switches or even make them piano key fashion out of sheet metal screws and 0.005-inch brass shim stock, the best bet is to purchase a surplus keyboard from another machine. These are available from a variety of sources (such as the classified ads in electronics magazines) for $15 to $30. Metal framed units without PC boards are the best, since the keys are easily rewired and rearranged. Try not to get an encoded unit because you'll probably have to pay extra for electronics you won't use. If the switches are mounted only on a PC board, cut the lands with an X-acto knife rather than dismounting all switches and fabricating your own frame. Most surplus units will also contain more than 40 keys. Retain or remove the extras at your option. You may want to use them for other functions (REPEAT, RESET, etc.).

One of your most vexing problems will be the labels for your new keyboard. For this, there's no easy solution. You can alter some keys with a fine-tipped engraver, but in most cases you will have to carefully print your labels on paper or plastic and then cut them out and glue them to the keys. Keys with removable caps and labels are available, but not on the surplus market.

Complete keyboards and keyboard kits specifically for the ZX81/TS 1000 are available, most with a repeat function, for about $60 to $80.

So, having assembled and wired your keyboard, where do you attach it? Remove the bottom of the computer case. Figure 8.6 shows the lower left corner and bottom edge of the PC board. Note that the keyboard connectors are on the other side of the PC board and attached to the solder terminals labeled D8–D1 and KBD4–KBD0 in Figure 8.3. To attach your keyboard, use a fine-tipped, low-power soldering iron to connect wires to each solder terminal in the lower row you see in Figure 8.3. Use color-coded wires or mark them. (Be careful not to allow solder whiskers to short between terminals.) Then connect the opposite ends of the wires to the appropriate half-row or column of your keyboard as Figure 8.1 shows. It's convenient to use 16-conductor ribbon cable (you can use the extra 3 conductors for other functions) with a 16-pin DIP plug on the keyboard end. Use a matching 16-pin socket on the keyboard.

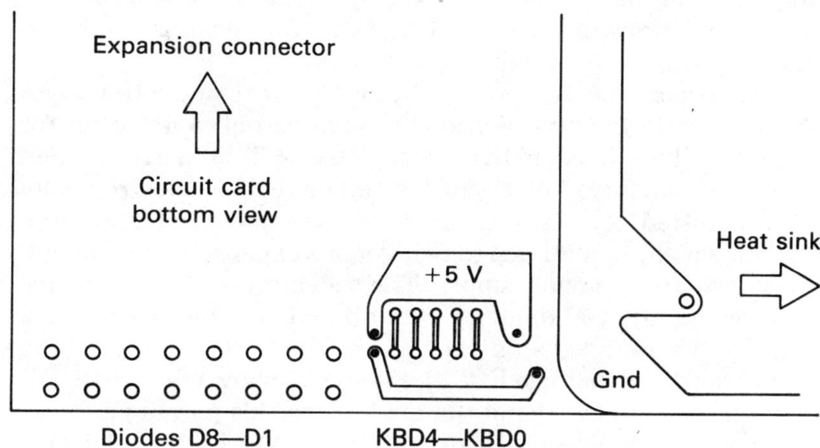

FIGURE 8.3 Keyboard Attachment

With a flat file, remove a section of the recessed lip of the top of the computer case below the letter "M." Make the section as wide as the ribbon cable, as Figure 8.4 shows. Feed the ribbon cable through this egress and secure it under the original keyboard with a piece of tape. Disconnect the original keyboard, taking care that its leads do not short against anything. Before reassembling the computer, test your keyboard.

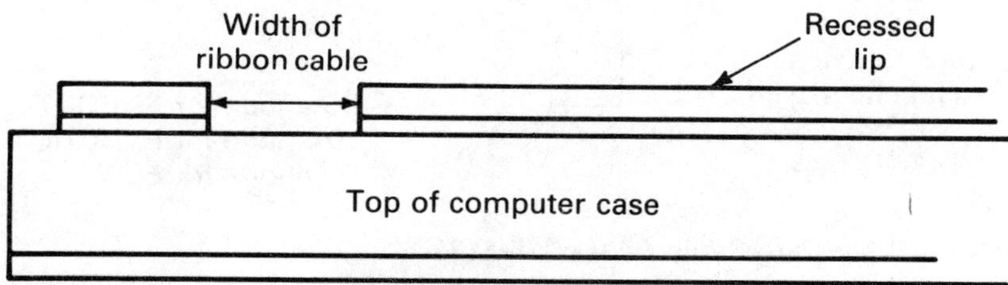

FIGURE 8.4 Cable Egress

If you plan additional modifications or experimentation, it will be handy to make a base for your keyboard that's also big enough for the PC board. You can make one by mounting the PC board on the base along with the keyboard and then wiring them together, discarding the original computer case and keyboard. Note, however, that running your computer without its case may cause unacceptable TV interference. If so, you can rectify it by placing a light metal or metallic window screen shield over and under the computer. Attach the shield to any of the ground points on the PC board.

Having come this far, you might also want to add a few extra convenience keys. These will allow you to use only a single key for the normally shifted **FUNCTION**, **RUBOUT**, and **EDIT** commands. While single-key operation can be realized electronically for any shifted key, the technique requires the use of integrated circuits which, in addition to being more expensive and complicated, require a power supply. The scheme you'll use requires only two germanium diodes per additional key, but it will work only for the four keys on the right- or left-hand edges of the keyboard (those that use KBD0). Figure 8.9 shows the circuit for each additional key. Using the diodes like this permits a 0-volt signal on address line A8/D6 or A12/D3 (or A14/D7 or A11/D1) to pull KBD0 low. This action is equivalent to pressing SHIFT (A8 pulls KBD0 low) and 0. (A12 pulls KBD0 low.) You can also do this for STOP, " ", ", or £ by attaching the second diode to the anode of D4, D2, D5, or D8, respectively—but most people don't feel much of a need to provide single-key operation for these functions. The 1N34A diodes can be purchased from Radio Shack in a packet of ten as part number 276-1123.

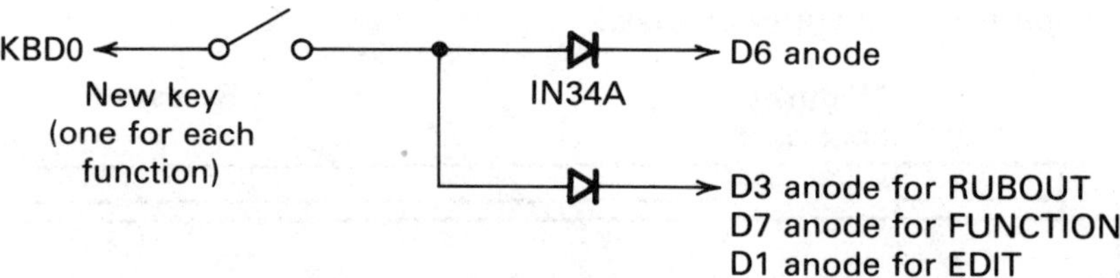

Note: Diodes are Radio Shack 276-1123.

FIGURE 8.5 RUBOUT, FUNCTION, and EDIT Keys

SHIFT-LOCK is another special function key that can be very useful, especially whenever you want to push the cursor around the screen or whenever you're doing graphics. It's a very easy function to realize. All you need do is obtain any common single-pole, push-on/push-off push-button switch and connect it in parallel with the SHIFT key. Pushing it on locks the SHIFT and pushing it off unlocks it. Try to find a key that is roughly the same size as your existing keyboard keys, but note that a slight difference in ''feel'' will make it more recognizable to the touch when you are not looking at your keyboard. If you want to get fancy, buy a double-pole switch and use the second pole to connect the 9.75-volt power supply to a light emitting diode (LED) through a 680-ohm resistor as a SHIFT-LOCK indicator. Or, if you've built a preregulator for your computer or a heftier 5-volt supply, you can connect the 5 volts to the LED through a 330-ohm resistor. Refer to Figure 8.6. If you scout around your local electronics stores, you'll probably find a switch that suits.

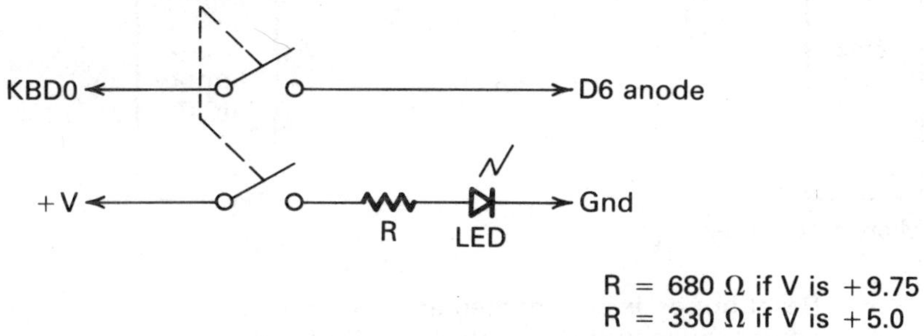

KBD0 ← ──o o──────────→ D6 anode

+V ← ──o o──〜〜〜──▷|──→ Gnd
           R   LED

R = 680 Ω if V is +9.75
R = 330 Ω if V is +5.0

Note: Key has double-pole/single-throw, push-on/push-off switch. (You can omit LED indicator circuit and use single-pole switch.)

FIGURE 8.6 SHIFT-LOCK Key

Who hasn't, at one time or another, hopelessly locked a computer into an endless loop or had the whole system crash? Do you get tired of crawling around the legs of your desk to pull the power supply out of the wall socket just to reset the machine? (By the way, don't accomplish the same thing by pulling the power plug out of the computer too often; this has a nasty habit of causing the spring contacts in the jack to lose their tension, causing intermittent power outages that wipe out the program you just spent half the day typing in.) Anyway, about the time you wear the knees out of your jeans you might start thinking about adding a RESET button to your computer. A RESET button accomplishes the same thing as pulling the plug without actually doing so, and it will even preserve anything you've stored above 32K or in the 8K–16K area of memory. (See Chapter 3.)

Installing the button is easy to do, but it does involve opening the computer and soldering to the top side of the PC board. See

Top View of Printed Circuit Board

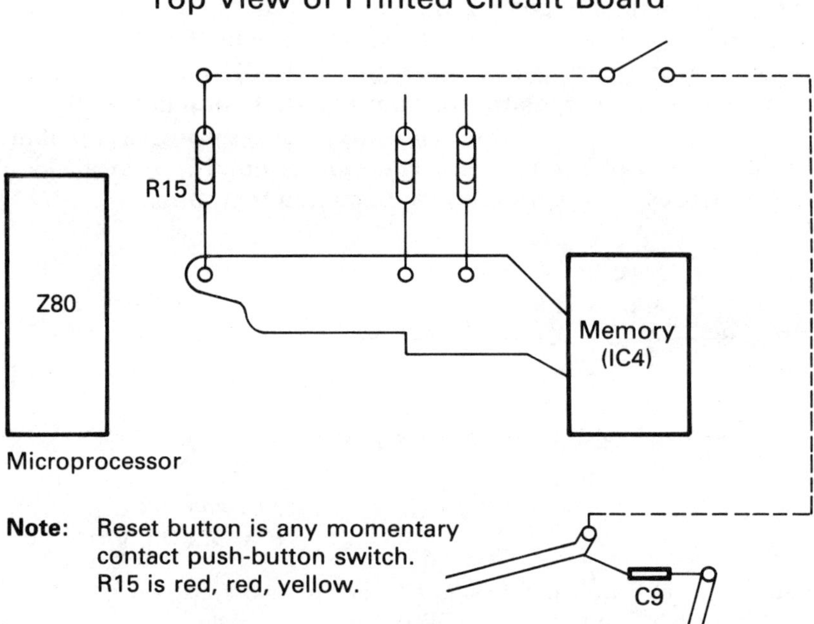

Note: Reset button is any momentary
contact push-button switch.
R15 is red, red, yellow.

FIGURE 8.7 RESET Button Wiring

Figure 8.7 for the complete wiring story. Any momentary contact push-button switch will work; a keyboard switch is fine. The only thing you must *not* do is mount that switch near the keyboard where randomly roving fingers can accidentally depress it. The rear top of the case, if you're still using it, is a good place. If you're not, you can put the key above or adjacent to the PC board.

The last special key is the most complicated. This is the REPEAT key. In principle, it's easy enough. Your computer scans the keyboard at the start of every TV frame. If a key is depressed, the computer accepts it. The key's value is stored in the "Debounce" location. On the next scan, if the key value the computer detects is the same as the value it previously stored, the computer will ignore it. This process prevents multiple letters from appearing accidentally because you couldn't get your finger off the key in a sixtieth of a second. In order to know that you've released the key, the computer must see either no key or a different key depressed. What you'll do, therefore, is build a circuit that periodically connects and disconnects the keyboard to the KBD lines whenever a REPEAT key is pressed. When it's disconnected, the computer will think that no key is depressed whether you're holding one down or not. When the circuit reconnects the keyboard, the computer will see whatever key you are still holding and accept it as a new entry. The circuit must also leave the keyboard permanently connected when you are not holding the REPEAT key so that it will function normally. The circuit could be used for auto-repeat by installing a jumper in place of the REPEAT key, but if you do, you'll sometimes miss a letter during normal typing because the keyboard will be disconnected when you hit the key. This may not be a problem if you're a hunt-and-pecker. Try it with a temporary jumper before you finalize your circuit to see which version you prefer.

You'll find the schematic diagram for the circuit in Figure 8.8. You may wish to change the values of CV or RV to change the rate at which characters are repeated. You should take the 5-volt supply and the ground for this circuit from the bottom of the PC board at the points shown in Figure 8.3. If you have some electronics experience, you're aware that you could use other ICs (*I*ntegrated *C*ircuits) and other circuits to achieve the same affect.

If you find a moving-key keyboard a big improvement in entering data on your TS 1000, you'll find a joystick an equally large improvement in game playing and other forms of cursor movement. To implement a five-function joystick you'll use only one half-row line and the KBD0–KBD4 lines. Which half-row you use is immaterial; for now, use 6–0 (A12/D3). You can have as many as eight five-function joysticks by using each of the half-row lines. You can attach the joystick either directly or via a six-wire connector of your choice. If you have an external keyboard, then it's generally easier to make your joystick connections to it rather than to the PC board.

The joybox in Figure 8.9 is made of Radio Shack parts, and it cost about $8, complete with mating connectors. You can lay out the switches to suit yourself, but putting four of them in a diamond pattern on the aluminum box top for up/down, right/left and the fifth for fire on the box end may be easiest to manipulate. The connection diagram for this arrangement appears in Figure 8.10.

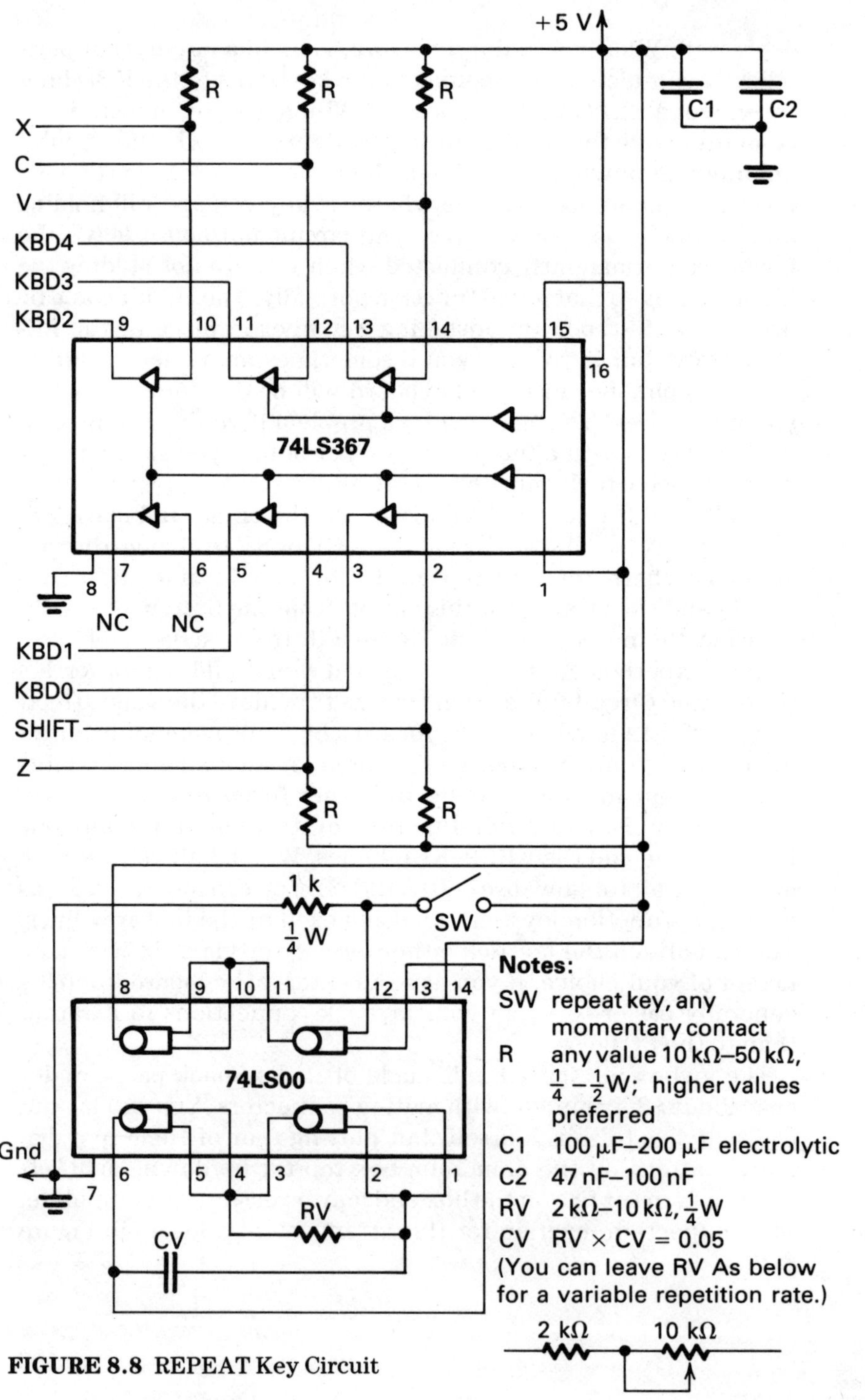

FIGURE 8.8 REPEAT Key Circuit

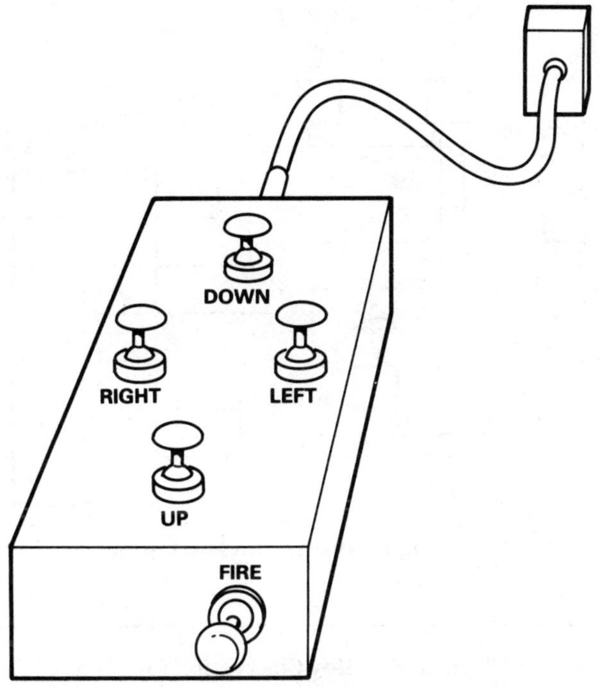

Notes:

Cable	8-conductor phone cable
Box	$3\frac{1}{4}'' \times 2\frac{1}{8}'' \times 1\frac{1}{8}''$ Radio Shack 270-23
Switches	5 miniature, push-button, normally open Radio Shack catalog number 274-1547
Connectors (optional)	6-pin, Radio Shack catalog numbers 274-207, 274-209

FIGURE 8.9 Joybox

You can use INKEY\$ to scan the joybox. If you use INKEY\$, then apply Table 8.1 to preserve the arrow conventions on keys 6–8. INKEY\$, however, will respond if and only if you press a single button. This means you cannot move and fire or move in more than one direction simultaneously.

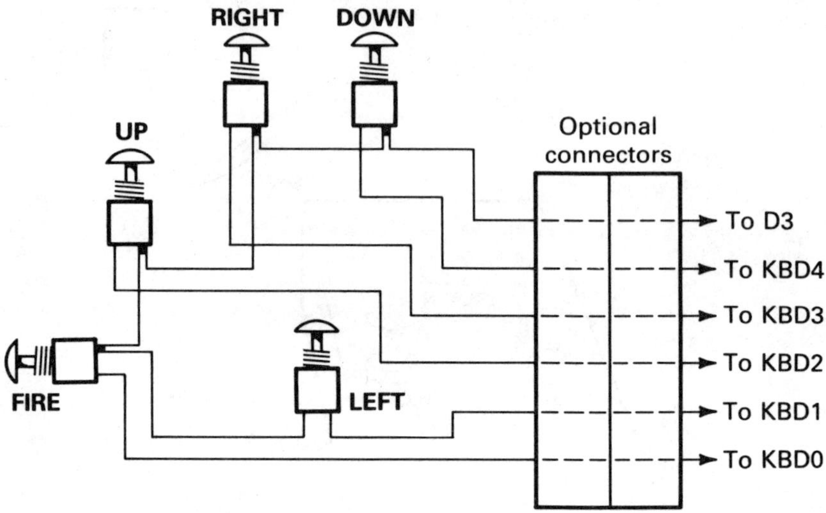

FIGURE 8.10 Joybox Wiring

TABLE 8.1 Key Conversion for Joybox

INKEY\$ VALUE	BUTTON FUNCTION
6	Down
7	Up
8	Right
9	Left
0	Fire

You can get around this problem by using machine code. To read
any combination of the five buttons in any half row use Program
8.1. These are the mnemonics.

Starting Location	Codes	Mnemonic
16514	62,a	LD A,a
16516	219,254	IN A,254
16518	79	LD C,A
16519	6,0	LD B,0
16521	201	RET

PROGRAM 8.1 Five-Key Reader

```
10 REM 12345678
20 LET X = USR 16514
30 PRINT X;"   ";
40 GOTO 20
```

Sequentially **POKE** in the following codes beginning at location 16514

62, a (see table below), 219, 254, 79, 6, 0, 201

Value of a	Half-row to which joybox is attached	Address line
127	B—SPACE	A15
191	H—ENTER	A14
223	Y—P	A13
239	6—0	A12
247	1—5	A11
251	Q—T	A10
253	A—G	A9
254	SHIFT—V	A8

What this routine does in the first four locations is read a value from the input port that corresponds to switch closures only from the half-row determined by a. This value is next placed in C while B is zeroed. The value of the BC register pair is then returned to BASIC as the decimal number X and printed. The program then repeats. Delete lines 30 and 40 when you are ready to build your game program, and put X to whatever use you wish. The value of X corresponds to the binary number

$$0 \quad 0 \quad 1 \quad SW1 \quad SW2 \quad SW3 \quad SW4 \quad SW5$$

where SWx = 1 if the switch is open and 0 if the switch is closed. With no switches closed, the number returned is 63. If, however, the joybox is attached to 6–0 (a = 239) and SW1 and SW3 (6 and 8) are pressed, then the number returned is

$$0 \ 0 \ 1 \ 0 \ 1 \ 0 \ 1 \ 1 = 43$$

Thirty-two different combinations are possible, although some of them, such as up and down simultaneously, don't make sense in most game situations.

At this point you could use IF statements to test for the various combinations of switch closures. For instance, if you wanted to take some particular action when 6 (down) and 9 (left) are pressed, then

xxxx IF X = 45 THEN (take action)

This method is slow and memory consuming (32 IF statements would be needed to test for all possible combinations!) but does allow a direct approach to the problem in BASIC.

Another approach is to decode the returned value so that you end up with five constants which are 1 or 0 depending on whether a switch is open or closed. (See Program 10.1.) However, you'll find that this technique is also too slow for most game applications. Short of doing your entire game in machine code, probably the best approach is to read and decode the keyboard in machine code and then return to BASIC. A method for doing this is demonstrated in Program 8.2. The program itself is designed to allow "blocky" pictures to be drawn anywhere on the entire 64×42 screen grid when you use the TS 1000. (The 1K machines don't have enough memory for the program and a full screen, but some drawings are possible.) You can run it using the keyboard (6–0) keys, but it's more fun with a joybox or joystick. Initially, a single pixel will be in the extreme upper right corner of the screen at coordinates 63,41. Attempting to plot above or to the right of this point will result in the program halting. It is wise at this juncture to use the down button (6 key) to form a border all the way down the right-hand edge of the screen. You will notice the line propagate down the edge and suddenly stop at the bottom. Release the down button and press the fire (0 key) and left (9 key) button. Notice that a blinking cursor leaves the border traveling left. Release the left button and press the up button (7 key). Notice that the cursor moves down! Keep it going until it bounces off the bottom; it will now be moving in the correct direction. Any time the cursor hits bottom the up/down keys will reverse and whenever it hits the left edge the right/left buttons reverse. Correcting them requires that the cursor bounce off the edge again. You've probably noticed by now that pressing the fire button allows you to move the blinking cursor wherever you would like while releasing it causes the points to be plotted. If you want to reverse the action change line 60 to

60 GOTO 70 + PEEK 16535

It is very difficult to draw circles manually, so use the circle plotting subroutine that begins at line 130. To access it, place the cursor where the center of the circle is to be and press the C key on the keyboard. The L-cursor will appear at the lower left of the screen. Enter the radius, but make sure it will not require that the circle be plotted outside of the 64×42 screen grid, or the program will stop. Try 2, 4, 8 to get a feel for how big the circle will be. Now, go eat lunch because that circle routine is slow, slow, slow. What you get doesn't look a whole lot like a circle, either, but it's the best the TS 1000 can do. There may be a hole or two in it along the X axis. You can eliminate these holes by decreasing the step size in line 140, though it will be at the expense of increased plot time. Actually, the holes are useful since you need a path to get the cursor out of the circle. You can fill them in as you leave. If you don't like or need the circles, then delete lines 85 and 130–200.

After you've completed your masterpiece you may want to save it. Connect your recorder, start it, and press the S key. The program and your drawing are saved as ''DRAW.'' When reloaded, your drawing will come on exactly as you left it, program still running. If you have no intention of saving the drawings, then delete line 90.

The mnemonics for this program are listed below.

Starting Location	Codes	Mnemonic
16514	33,151,64	LD HL, 16535
16517	6,5	LD B,5
16519	62,239	LD A,a (a=239)
16521	219,254	IN A,254
16523	47	CPL
16524	79	LD C,A
16525	62,1	LD A,1
16527	161	AND C
16528	119	LD (HL),A
16529	35	INC HL
16530	203,25	RR C
16532	16,–9	DJNZ –9
16534	201	RET
16535	0	DATA
16536	0	DATA
16537	0	DATA

| 16538 | 0 | DATA |
| 16539 | 0 | DATA |

What this routine does is to load the C register with the value corresponding to the keys which have been pressed in the half row determined by a (in this case, keys 6–0). This value is then shifted right five times and following each shift the least significant bit is compared to the value one. The result of each comparison is stored sequentially in locations 16535–16539 so that at the end of the routine they contain a value of one or zero corresponding to whether keys 0 through 6 were pressed or not. Other half rows can be used by altering the value of a according to the table given with Program 8.1.

PROGRAM 8.2 Draw

```
10 REM (26 characters)
20 LET X = 63
30 LET Y = 41
40 PLOT X,Y
50 RAND USR 16514
60 GOTO 71 – PEEK 16535
70 UNPLOT X,Y
71 LET X = X + PEEK 16537 – PEEK 16536
80 LET Y = Y + PEEK 16538 – PEEK 16539
90 IF INKEY$ = ''C'' THEN GOSUB 130
100 IF INKEY$ = ''S'' THEN SAVE ''DRAW''
110 GOTO 40
130 INPUT R
140 FOR J = 0 TO 2*R STEP .2
150 LET Z = INT (.5 + SQR (2*R*J – **2))
160 LET Q = INT (.5 + X – R + J)
170 PLOT Q,Y + Z
180 PLOT Q,Y – Z
190 NEXT J
200 RETURN
```

Sequentially **POKE** in the following codes beginning at location 16514

33, 151, 64, 6, 5, 62, 239, 219, 254, 47, 79, 62, 1, 161, 119, 35, 203, 25, 16, –9, 201, 0, 0, 0, 0, 0

Chapter 9

Power That Doesn't Corrupt

Politico-philosophical considerations aside, a clean, stable power supply is essential for the proper operation of your computer. Corrupting influences include current limiting, heat build-up, noise, voltages that are too high or too low, and 60-Hz line components that are not sufficiently filtered out. Assuming that the plug-in power supply that comes with your computer is in working order and that the regulator on the PC (*Printed Circuit*) board is not defective, the only problem you're likely to encounter is heating in the computer. If you add no extra memory or other circuits to your machine, you probably needn't worry about this. However, most of the add-ons you can buy for the TS 1000 draw their power from the on-board computer supply, and all the circuits in this book can be powered from it.

To understand why the computer heats up you must examine how the power is generated. The unit you plug into the wall supplies 9.75 volts DC (*Direct Current*) at up to 600 milliamps (ma). These are nominal values; the voltage will be somewhat higher with no load and should approach 9.75 volts as the load increases to 600 ma. The amount of AC (*Alternating Current*) mixed with it will also increase with the load. Now the components of the computer itself will function properly only at 5 volts DC and will draw a little over 300 ma at that voltage. The 300 ma current is fine. It's only half what the wall unit can supply, so you don't have to worry about current limiting. But that voltage is way too high. If you open your computer, you'll see a flat aluminum plate attached with a screw to a 3-legged IC (*Integrated Circuit*) on the PC board. The aluminum plate is a heat sink designed to dissipate the heat generated in the IC. The IC is a 7805 positive 5-volt regulator. Its job is to knock the 9.75 volts down to 5 volts and keep it within a few millivolts of that value. It is capable of controlling up to 1500 ma provided it doesn't get too hot. The reason it gets hot is the extra 4.75 volts. Power is defined as voltage times current, so the power the 7805 dissipates is the 4.75 volts across it times the 300 ma through it, or 1.425 watts. The device is specified as being able to dissipate 2 watts in free air at 20 degrees centigrade. Well, it does have a heat sink, but it's not in free air. Let's say that one cancels the other and that, as installed, it can dissipate 2 watts. If you still drop 4.75 volts across it, then the most current it can handle is 2/4.75 = 421 ma. So, you have a 100 ma surplus capability *at most*.

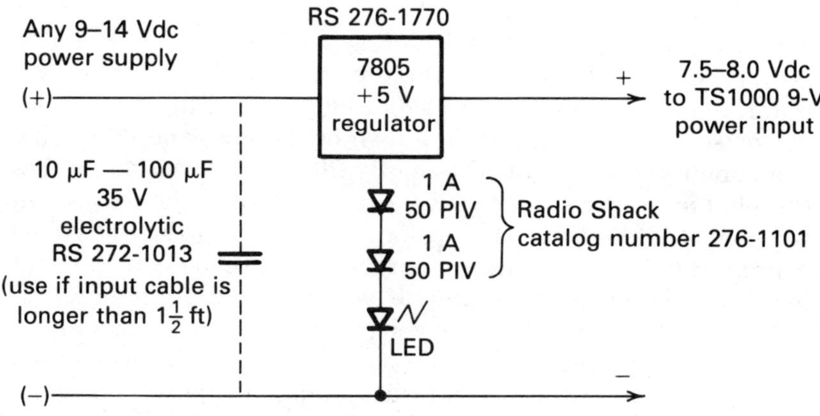

Note: Attach heat sink to 7805.
Either insulate heat sink with
mica washer or do not ground it.

FIGURE 9.1 Preregulator for the TS 1000

If you plan any additions that total more current than this, you're going to have to make some changes. You could add a second heat sink (or a larger one) and paint both it and the original flat black, cut air holes in the computer case (which might weaken it), add a fan, or do any combination of these things. But you can also solve the problem electronically—and solve it more cheaply and easily—by building a preregulator. A preregulator is nothing more than another 7805, a couple of diodes, and some household scraps. The total cost is about $3. You'll find the schematic for it in Figure 9.1. You can build it in a box to set next to the TS 1000, attach it to the wall unit, or assemble it "in-line." You must not ground the heat sink or you'll reduce the output voltage to 5 volts. If you have a voltmeter, connect it to the output of the pre-regulator and try various diodes until the output voltage is as close to 7.5 volts as you can get it. If you don't have a voltmeter, then use the diodes specified; they will give you about an 8-volt output. One of the diodes is an LED so that it can also act as a pilot light. What this circuit does is reduce the voltage drop across the on-board regulator to 2.5 or 3 volts. It can now dissipate 667 ma, which is more than the wall unit can safely supply, and you now have a 300 ma surplus. Unfortunately, if you have a Timex/Sinclair 16K memory pack you cannot use this scheme because the pack requires the 9.75 volts supplied by the wall unit. Memotech 64K memory packs will work with the preregulator. Other brands may as well.

You'll need a 5-volt supply for the projects in this book. If you've built the preregulator it's all right to take it from the PC board at the points shown in Figure 8.6. If you haven't built it or don't want to open your computer, then you have several other options. The easiest is to get four NiCad batteries and a battery holder. They will supply almost exactly 5 volts, and you can also get a charger to recharge them. Any cells from AA to D will work. The smaller ones will have to be recharged more often, of course. *Do not* use carbon-zinc or alkaline cells. Four of them will supply over 6 volts, and that's too much voltage for some of the chips you'll be using.

A second choice is a separate 7805. The schematic for this power supply appears in Figure 9.2. You can use any 7.5- to 15-volt DC supply, regulated or unregulated. The 10-volt, 120-ma charger for a Black & Decker battery-operated drill works great, for example. But remember

$$\text{maximum current} = 1000 * 2/(\text{supply volts} - 5) \text{ ma}$$

If you are not pulling anything like 600 ma from the wall unit, you can use it! Buy a 2-to-1 1/8-inch phone plug adapter, and use it as the drawing in Figure 9.2 shows.

If you're a little more ambitious or if you're planning some larger projects in the future, you might want to build the following power supply. It can provide up to 2 amp total current and, by

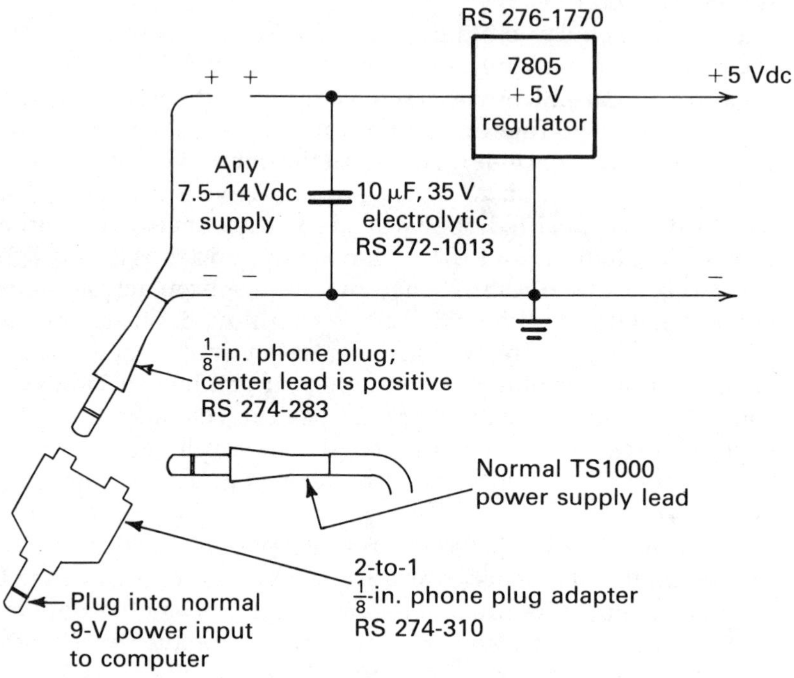

FIGURE 9.2 Five-Volt Experimenter's Supply

attaching your preregulator and 5-volt supplies to its output, will provide 9 to 14 volts (decreasing with load) unregulated and 5 and 8 volts regulated current. You can use the unregulated output with most CMOS circuits and the 5-volt for TTL (*Transistor Transistor Logic*) devices. If you add the preregulator, you can put the wall unit supplied with your computer on the shelf. The power supply and 5-volt regulator in Figure 9.3 are on a $12'' \times 3^{1/2}'' \times ^{1/16}''$ piece of aluminum that acts both as a heat sink for the regulator and as the rear legs for the particle board base on which the computer PC board and keyboard are mounted. The 7805 preregulator is mounted on a separate $5'' \times 5''$ piece of aluminum and screwed to the underside of the particle board. A small terminal board is handy for making connections. Figure 9.4 gives the schematic.

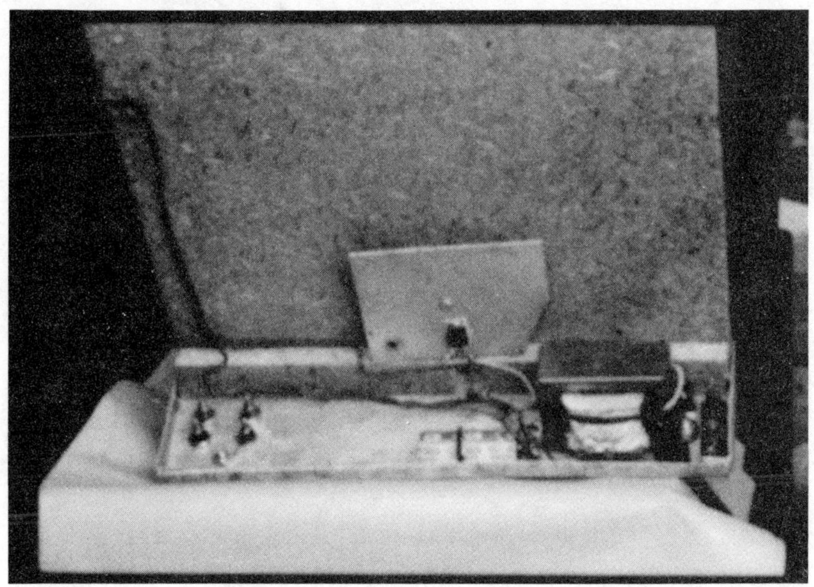

FIGURE 9.3 Underside of Particle Board Base Showing Power Supply and Preregulator

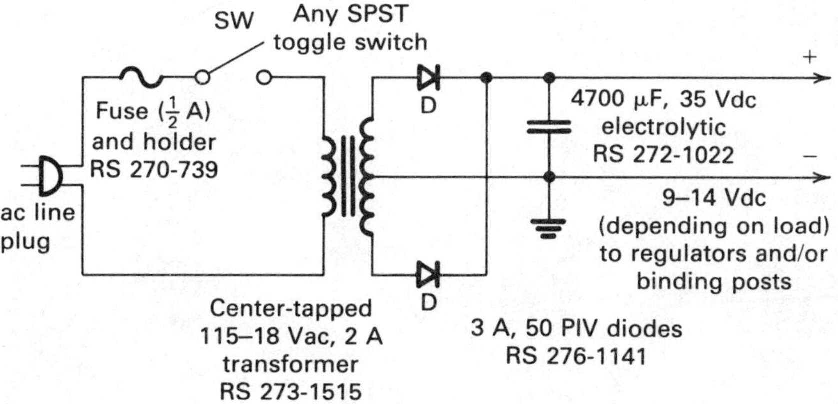

Note: Pilot light can be added to input or output of transformer. Secondary 2-A fuse can be added in (+) line after capacitor.

FIGURE 9.4 Two-Amp Power Supply

Chapter 10

Sensing the Outside World

Almost invariably, someone who becomes reasonably proficient at programming soon begins to wonder if he or she can't make a computer ''do something.'' Usually this translates into ''Isn't there some way that my computer can affect or be affected by the outside world other than by my typing on the keyboard? Can't it tell when it's dark? hot? wet? whether the garage door is open? or any of a number of other things? Won't it turn on the lights or close the garage door?'' The answer is a qualified yes, which means certainly, but usually not without some peripheral equipment. To ''do something'' the first items you need are input/output (I/O) ports. These are special-purpose circuits that allow outside signals into the computer and inside signals out of it under program control. When it comes to sensing physical parameters, you'll also need a transducer, a device that transforms a physical parameter, such as temperature, into an electrical signal. Still another circuit, called an interface, is often essential. This converts the electrical signal from the transducer into one that's compatible with the input circuit of the computer. (In some cases, you can eliminate it.) You can purchase these devices ready-made and connect them to your computer fairly easily, but their price rivals that of the computer itself.

To see what you can economically make the TS 1000 do, first take a look at how it might sense the outside world. Like most TS 1000 owners, you may not realize that there is what amounts to a 6-bit input port hidden inside that black plastic box. That is, there is a circuit that will accept 6 different on/off signals from outside. One is obvious; you use it whenever you load a taped program. It's unfortunate, but the ear jack (labeled EAR) is AC-coupled to the outside world and, in general, without changing the PC (*Printed Circuit*) board the jack will accept only alternating signals, not DC (*Direct Current*). The other five are less obvious. Remember those KBD0–KBD4 lines from the chapter on keyboards? You guessed it. In fact, outside of the interfacing on the PC board, all 6 inputs (which are actually contained in the ULA chip) are identical. The KBD lines, as you found out earlier, are held up to 5 volts by 10K resistors, while the EAR input is held down to ground and coupled to the ULA through a capacitor. Figure 10.1 is a block diagram of the type of sensing systems you can assemble with the TS 1000.

So how does this help you sense the outside world? Go back and read the discussion concerning Figure 8.2. Notice that you can substitute any other switch for a keyboard key; this is how you built the joybox. Want to know if the door is open? Put a switch on the door jamb that closes whenever the door opens and run the leads to one of your keys.

Although this technique works in principle, it suffers a couple of shortcomings. First, if the switch is closed while you're trying to program, the computer will interpret it as a key being held

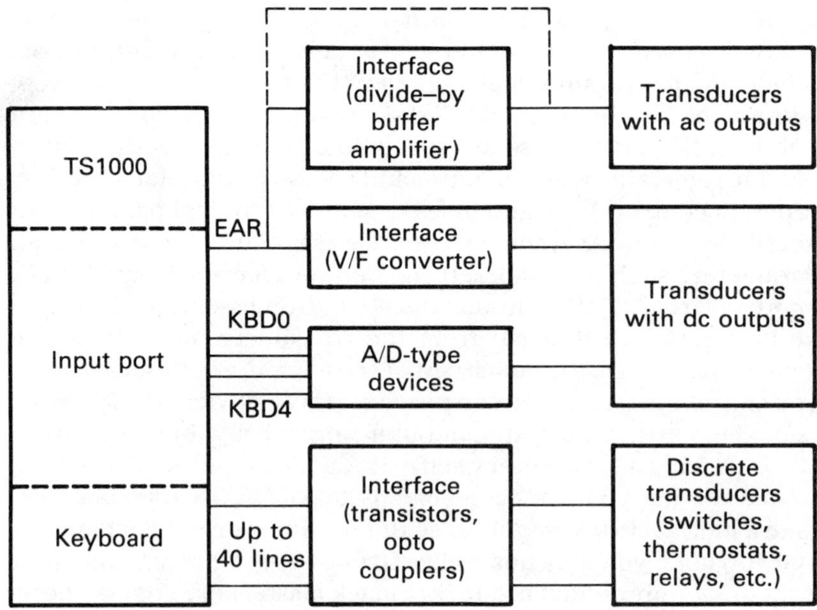

FIGURE 10.1 Block Diagram of TS 1000 Sensing Systems

down and refuse to accept any other keystrokes until it is released. You can get around this easily by not attaching the wires to the keyboard until after your program is running. You just connect and disconnect the switches with a connector or put a simple on/off switch in series with the incoming switches. For instance, you might type in the following simple program:

```
10 IF INKEY$ = "W" THEN PRINT AT 0,0;THE DOOR IS
   OPEN"
20 IF INKEY$ = " " THEN CLS
30 GOTO 10
```

Start the program running and then attach your door switch (or simulated door switch) across the W key. When the door switch closes, the computer will print "THE DOOR IS OPEN," and when the switch opens, the screen will clear. You can use this technique to sense anything you can put a switch on—or anything that already has a spare switch closure. You may attach the switch to any key except BREAK. Want to know when the temperature reaches 75 degrees? Find a cheap thermostat, and attach the contacts across a key. Set it to 75 and use a program similar to the one you just tried. If you'd like an audible alarm, try this.

```
10 REM (machine code tone generator, Program 7.2)
20 IF INKEY$ = "(KEY OF CHOICE)" THEN RAND USR
   16514
30 GOTO 10
```

The parentheses mean that you should do what's inside of them rather than typing the contents literally. With this program, the alarm will sound when the temperature rises above 75 (or any other temperature you set your thermostat for). You'll have to reset it with the BREAK key.

Another problem with this scheme is that it may not work if the wire length is more than a few feet. Any pair of wires acts as a capacitor; the longer the wires, the greater the capacitance. At the pulse widths used to scan the keyboard, it doesn't take much of a capacitor to start looking like a short across the key. A secondary consideration is that cables also pick up noise. The noise can be anything from a 60-Hz hum to interference from your local TV station if you're close enough. But try it anyhow. If it doesn't work, then read on. Switches sometimes work on lines as long as 40 feet.

If you simply must run long lines, there's a cheap way to do it. In fact, this method is preferred even for short lines because it completely isolates the computer electrically from whatever it is you are trying to sense, thereby protecting the machine from unintentional damage. You will use an opto-coupler, a small integrated circuit containing a phototransistor and a light emitting diode (LED), which are not electrically connected. A current of about 20 ma (milliamps) passes through the diode, causing it to emit light that the phototransistor receives, causing it to turn on. If you connect the collector of the NPN phototransistor to one of the KBD lines and connect the emitter to an address line (i.e., across a keyboard key), the transistor will effectively short them together when it turns on. Figure 10.2 shows the simple schematic for a typical circuit. Since you are sensing a DC signal, the lines can be virtually as long as you desire, although the cable resistance cannot limit the current to less than 20 ma. The external diode prevents damage to the chip should you be using a high voltage and connect it backwards. If you're careful, you don't need this diode, but if you decide to use it, any 1N400x series diode will do. The capacitor and 100K resistor are used optionally with very long lines to provide a degree of noise immunity; you can omit them for lines shorter than 100 feet in most environments. The resistor, R, should be approximately $(V - .7)/.02$ ohms. Exact values are not needed; $R = 390$ ohms for $V = 5$ volts works quite well.

Still another problem with sensing via the keyboard is that you can look only at one key using the INKEY\$ function. You could get

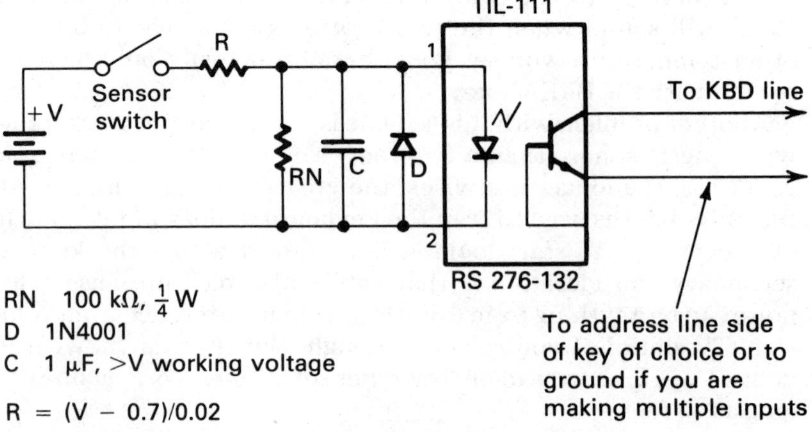

FIGURE 10.2 Typical Opto-Coupler Circuit

around this limitation by writing your own machine code keyboard scanning routine as in Program 8.1 or even using the one in the TS 1000 ROM (*Read-Only Memory*). You'd then be able to attach a switch to each key except BREAK. What you'll do instead is write a very short routine that will let you look at any of four inputs. This will satisfy most users, and if your needs are more ambitious than that, you'd better learn some machine language and write your own program. Type in the following lines.

```
10 REM 123456
20 LET X = USR 16514
30 PRINT AT 10,10;X
40 GOTO 20
```

POKE in the codes listed below with the mnemonics.

Starting Location	Codes	Mnemonics
16514	219,254	IN A,254
16516	79	LD C,A
16517	6,0	LD B,0
16519	201	RET

Start the program **RUN**ning with your switches/opto-couplers disconnected from the keyboard. The number 63 should appear near the center of the screen. Now connect the collectors of the opto-couplers or one side of the switches to KBD1 through KBD4. Do not connect KBD0—it may cause a break. Connect the emitters of the opto-couplers or the other side of the switches to ground rather than to the address lines. This approach does not require that any of the address lines A8–A15 be made zero; the keyboard is not scanned. It will accept any grounds on KBD0–KBD4 as inputs, but a ground on KBD0 will be seen as a BREAK upon return to BASIC. What prints on the screen is the decimal equivalent of a binary number given by

    0 0 1 KBD4 KBD3 KBD2 KBD1 1

where KBDi is a 1 if the switch is open or the phototransistor is turned off. For all of them off, the number is 00111111 or 63. If only KBD1 is turned on, the number is 61; KBD4 only, 47, and so on. There are 16 possibilities.

There are several ways to decode these numbers. The following additional program lines will give you a value of $Ki = 1$ if KBDi is 1.

```
30 LET X = X – 33
40 LET K4 = INT (X/16)
50 LET K3 = INT (–2 ∗ K4 + X/8)
60 LET K2 = INT (–4 ∗ K4 – 2 ∗ K3 + X/4)
70 LET K1 = INT (–8 ∗ K4 – 4 ∗ K3 – 2 ∗ K2 + X/2)
80 PRINT AT 9,0;K4,K3,K2,K1
90 GOTO 20
```

You can now have the computer take whatever action you desire based on the condition of any of the switches. This approach, by the way, actually makes use of the input port as an input port rather than as a keyboard scanner.

Program 10.1 lists the finished routine including the **PRINT** statement necessary to let you watch the opening and closing of the switches on the screen. You can, of course, delete lines 80 and 90 and use the values of K1–K4 for whatever purpose you have in mind.

Can you see how to extend the program to decode all 5 KBD lines? (Hint: change 33 to 32, and add a new line, 75, which continues the number sequence in the argument of the INT functions for another iteration.)

Suppose that instead of sensing an on/off condition, you want to detect the level of a signal. So long as you are content to divide the signal range into no more than five segments, you can do this conveniently with a single LM3914 dot/bar display driver. This IC contains 10 voltage comparators connected to 10 LED drivers. One input of each comparator is fed from a resistive voltage divider, which can be supplied by either an on-chip voltage reference or an external voltage.

PROGRAM 10.1 Four-Input Reader and Decoder

```
10 REM 123456
20 LET X = USR 16514
30 LET X = X – 33
40 LET K4 = INT (X/16)
50 LET K3 = INT (–2 ∗ K4 + X/8)
60 LET K2 = INT (–8 ∗ K4 – 4 ∗ K3 – 2 ∗ K2 + X/2)
80 PRINT AT 9,0;K4,K3,K2,K1
90 GOTO 20
```

Sequentially **POKE** in the following codes beginning at location 16514

219, 254, 79, 6, 0, 201

Each step on the divider is one-tenth of the voltage across it. The external voltage can vary between 0 and 5 volts and can be offset from 0. There are ten outputs, but you will only use four of them so that you can read them in exactly the same fashion as you did for the opto-couplers. You can choose whichever outputs suit your needs. If you use the second, fourth, sixth, and eighth, you will have divided the signal input range into five equal parts; less than 1/5 the reference, between 1/5 and 2/5, 2/5 and 3/5, 3/5 and 4/5, and greater than 4/5. On the other hand, if you knew the signal were going to be very near a particular value, you might want to select four adjacent outputs bracketing that value.

This might be handy in building a flashlight battery tester because you know that a good battery will have a value very near 1.5 volts. What you're doing is using this device as a crude analog-to-digital (A/D) converter with the help of the computer. A true 4-bit A/D converter resulting in 16 voltage levels could be used in the same way, but you should probably put that project off until you know a little more electronics. If you already understand these devices, simply hook the four output lines (or the four most significant bit lines of your A/D if it's more than a 4-bit unit) of a free-running 4-bit A/D to KBD1–KBD4.

The schematic diagram for four-level voltage input appears in Figure 10.3. the LEDs and resistors are not necessary. They're included only so that you can see which output is active when testing the device. You can remove them after testing if you wish. The chip is being run in the bar mode, so more than one output may be active at a time. If you use the dot mode, data may be lost unless you also use contiguous outputs. The potentiometer supplies a variable input for testing. Make sure you have a ground connection between the computer and the chip. Run the same program you ran for the opto-couplers, and plug in the KBD lines. As shown, the KBD lines will go low sequentially (Ki's will be zero) for inputs of 0.2, 0.44, 0.7, and 0.95 volts. For a 5-volt input range, disconnect pin 6 from pin 7, and connect pin 6 to pin 3. The input switch points will now be 0.8, 1.8, 2.8, and 3.8 volts. See the data sheet that somes with the IC for more information on changing ranges, offsetting, and other items.

You've discovered a few of the things you can do with the KBD lines. Now let's see what you can do through the EAR jack. As mentioned earlier, it wants to see an AC signal under most circumstances but will accept down to DC if the input port is gated on and off very rapidly. However, it is AC signals from about 1 Hz (hertz)

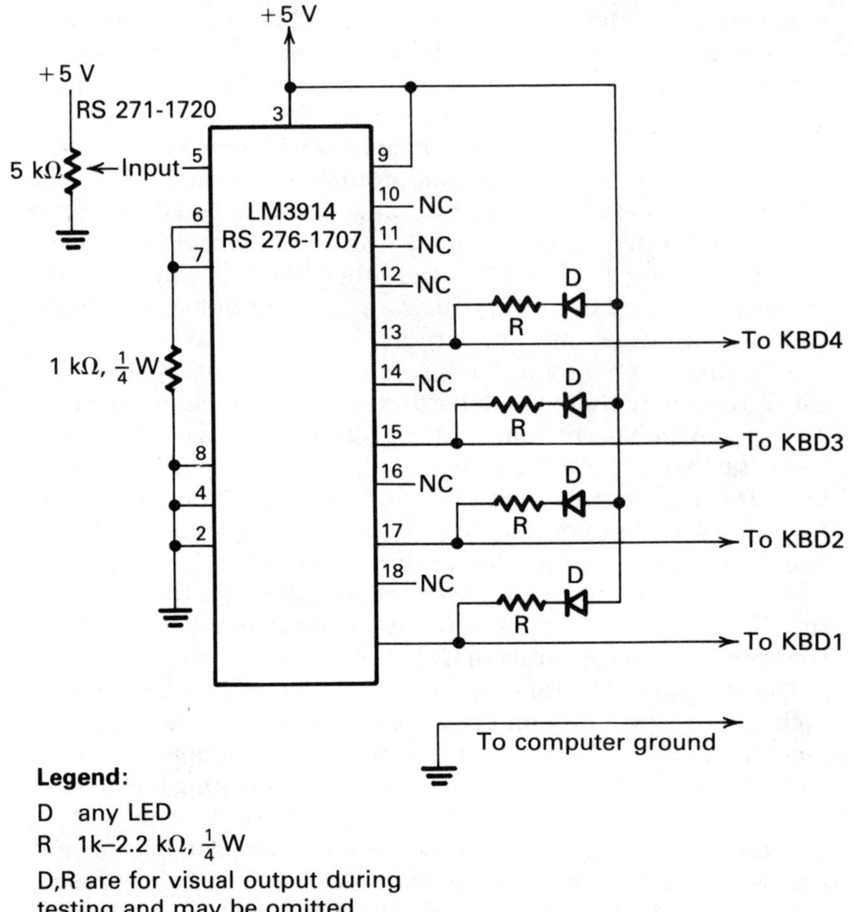

Legend:
D   any LED
R   1k–2.2 kΩ, $\frac{1}{4}$ W
D,R are for visual output during
testing and may be omitted.
5kΩ potentiometer supplies sample
input. It is not part of circuit.

FIGURE 10.3 Four-Level Voltage Input

to several MHz (megahertz) that you'll concern yourself with. Perhaps the most notable characteristic of AC signals is that they change from plus to minus and back again repeatedly. For periodic signals (those that repeat themselves in a definite interval), the most important parameter is probably their frequency, or how many times they change back and forth in one second. A plucked string on an instrument, for instance, vibrates at a specific frequency that we perceive as a particular tone. The unit of measure is the hertz, which is one change or cycle per second. Human hearing is most acute near 1 KHz (kilohertz) and diminishes as the frequency nears 20 Hz or 20 KHz in young people. The limits become somewhat tighter as you age.

The instrument used to measure frequency is the frequency counter, or simply, the counter. We are going to turn the TS 1000 into a counter. "Wait," you say. "What do I want a frequency counter for?" Well, you might want to know the frequency of your guitar strings or you might just want to find out how far your hearing extends. You might also want to build some of the remaining projects in this book, and the counter is the basis of some of the most important ones.

Type in Program 10.2.

The mnemonics follow.

PROGRAM 10.2 Frequency Counter

```
10 REM (25 characters)
20 FAST
30 LET X = USR 16514
40 LET F = INT (.5 + (X/(65535 * 43 +
   8 * X)) * 3.23 * 1000000)
50 PRINT AT 11,10;F/1000;" KHZ"
60 SLOW
70 FOR I = 1 TO 50
80 NEXT I
90 CLS
100 GOTO 20
```

Sequentially **POKE** in the following codes beginning at location 16514

17, 1, 0, 33, 0, 0, 68, 77, 219, 254, 25, 216, 23, 48, −7, 3, 219, 254, 25, 216, 23, 56, −7, 24, −17

Starting Location	Code	Mnemonic
16514	17,1,0	LD DE, 0001
16517	33,0,0	LD HL, 0000
16520	68	LD B, H
16521	77	LD C, L
16522	219,254	IN A, (FE)
16524	25	ADD HL, DE

16525	216	RET C
16526	23	RLA
16527	48,–7	JR NC, –7
16529	3	INC BC
16530	219,254	IN A, (FE)
16532	25	ADD HL, DE
16533	216	RET C
16534	23	RLA
16535	56,–7	JR C, –7
16537	24,–17	JR –17

The machine code routine comes in three parts. The first part (16514–16521) initializes the HL, DE, and BC register pairs. The second (16522–16528) is a loop that continues until the EAR input goes positive. On each pass through the loop the HL pair is incremented (16524). When the input goes positive, the BC pair is incremented (16529) and control passes to the third part (16530–16536), which is a loop that waits for the input to go negative, also incrementing the HL pair on each pass. Control then reverts (16537, 16538) to the first loop. This continues until the HL pair is filled up (65535 loops). Control then returns to BASIC (16523, 16533), and the contents of the BC pair show up as X. The algorithm in line 25 computes the integer value of the number of times the signal went positive divided by the time to complete 65535 loops. Line 30 converts the answer to KHz and prints it. The machine code executes in FAST mode while the display takes place in SLOW. Changing the upper limit on I in line 40 changes the amount of time the display is on the screen before the next measurement is taken. The signal into the EAR jack must not exceed plus or minus 5 volts but must rise above 2.6 volts for at least 15 microseconds and fall below 0.8 volts for at least 15 microseconds. Note that as long as these conditions are met no interface is needed. The resolution of the counter is 1 Hz from 0 to 33 KHz. The accuracy is plus or minus 0.5 count plus or minus 0.06% for the variations in my clock frequency. In other words, you could have a maximum error of 20 Hz when counting a 33-KHz signal or a 1-Hz error when counting a 10-Hz signal though usually the errors will be smaller. You could also average the value of F over several counts to reduce the error.

To measure the frequency of your guitar strings or any other audible tone, connect a microphone to the input of your amplifier and its output to the EAR jack. With the counter running, produce a continuous tone at the microphone (by repeatedly strumming if necessary) and increase the amplifier volume control. With the Radio Shack unit, you can safely leave the volume at maximum.

You can easily use this counter as a tachometer. Connect a reed switch between the +5-volt supply and the hot side of the EAR input. Use leads at least several feet long. Attach a small bar magnet to the spinning shaft, wheel, etc., that you want to measure. Move the reed switch close to the spinning magnet, and start the counter counting. Change F/1000 to F*60 and change KHz to RPM in line 30, and you're all set.

If you want to use your counter as a general-purpose test instrument at audio frequencies only, you should build a buffer (a single-chip Schmitt trigger will do nicely) or be very careful not to exceed the voltage limits listed in this chapter, or you'll risk damaging your computer. In many cases, however, you'll want to be able to measure higher frequencies. To do this you'll use an external divide-by-N device, which will also act as a buffer. You can use decade, binary, or divide-by-12 counters from the 7490, 7493, or 7492 families of chips. By using two of them cascaded you can achieve divisions of from 2 to 256. A 1-Hz to 8-MHz counter is a powerful test instrument. Check the parameters of the particular chip you buy, since the older 74L9x series will not count much above 3 MHz, so dividing by anything above 100 would be meaningless. The newer 74LS9x and 749xA series are good up to 16 MHz. If you wish to count even higher, you will have to obtain one of the special high-frequency counter prescalers often advertised in electronics magazines. This book can't begin to show you all the possible variations you can use to obtain different divisions with all thse chips; you'll have to consult their applications notes. But Figure 10.4 shows a divide-by-10 circuit based on a 74LS90 chip. Radio Shack's *Engineer's Notebook* also contains a number of divide-by circuits using these chips. Almost all require only the chip itself and differ only in pin connections. To obtain higher divisions, connect the output of a divide-by-N1 circuit to the input of a divide-by-N2 circuit. The second output will be a divide-by-N1*N2.

If you divide by 10 you will have to alter the scaling in your program because the true frequency is 10 times higher than that the computer counts. Change F/1000 in line 30 to F/100. The resolution is now 10 Hz, and the accuracy is plus or minus 5 counts plus or minus 0.06%.

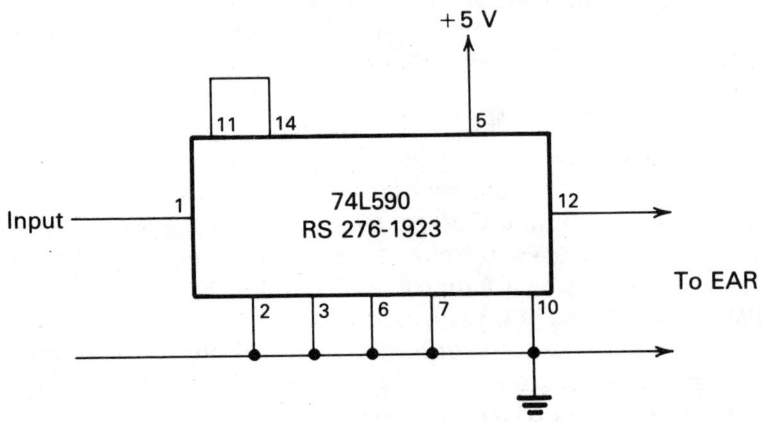

FIGURE 10.4 Divide-by-10 Circuit Using 74LS90 Chip

You can, of course, use this program/circuit strictly as a frequency counter, which is a handy instrument to have if you are prone to playing around with electronics. You might also find applications for it if you are a musician or audiophile, but if you want to measure physical parameters of the real world, you'll have to convert it to a voltmeter, primarily because most transducers you can buy or make have a constant voltage output for a fixed input. For now, forget about the divide-by and substitute a 9400 voltage-to-frequency converter chip as your interface. It converts a voltage, which is not suitable for inputting to the EAR jack, to a periodic AC signal that is.

The 9400 produces a pulse output whose frequency is linearly proportional to its voltage input. (Wired differently it can also produce a voltage proportional to the frequency of an input signal, but you don't need that capability here.) Since your program measures frequency, all you have to do is determine the voltage-to-frequency proportionality constant and appropriately scale your frequency measurement.

The circuit diagram is shown in Figure 10.5. Normally the output transistor of this chip is wired for common emitter, open collector operation, but because the input of the EAR jack is shunted by a 220-ohm resistor, it's configured as an emitter follower. So don't be disturbed that the output isn't wired according to the instruction sheet that comes with the chip. The 9400 is rated for operation with 8- to 14-volt supplies and is dependent on a well-regulated reference voltage for accuracy. Rather than use two separate supplies, you can try using the +5-volt regulated computer supply for both. The circuit appears to function perfectly at that voltage. If there is any degradation of linearity, you'll need a good digital voltmeter to detect it. You can change the 22K resistors to any value from 5K–33K without affecting the operation of the circuit as long as both resistors are equal. You can also change the capacitors somewhat, but read the instruction sheet before doing so. The 1K potentiometer supplies a variable

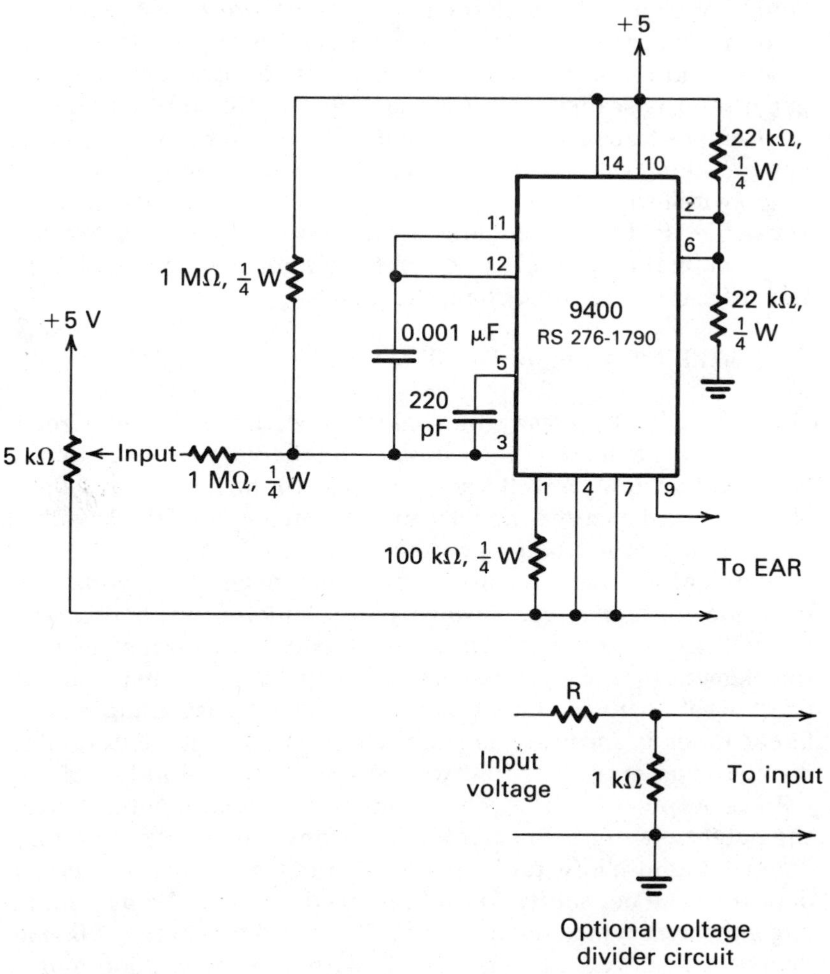

FIGURE 10.5 Voltage-to-Frequency Converter Using a 9400 Chip

voltage for testing purposes and you should omit it when your circuit is complete. For a circuit like the one in Figure 10.5, the output frequency the counter measured was 0–3963 Hz for 0–5 volts of input. You can increase the sensitivity by reducing the value of both 1M resistors, but this will also reduce the input impedance of the voltmeter. For higher voltage ranges, attach the optional voltage divider to the input. If V is the maximum voltage you wish to measure, the value of R will be $(V - 5)/5$ in kilohms.

If you have some electronics experience, you'll realize that you could attach a rectifier to the input to measure AC voltages or a shunt resistor to measure currents. In fact, with a rotary switch, diodes, and resistors you could turn this circuit/program into a general-purpose digital multimeter. Add the divide-by circuit and another position on the switch for a powerful counter/multimeter combination. Don't forget to leave enough poles/positions on your switch so that you can supply grounds to the KBD lines and cause the TS 1000 to change scales automatically as you switch.

To obtain the correct reading for the circuit as shown, and to get it on the screen in volts, change line 30 to

```
30 PRINT AT 11,10;INT (1.26 * F)/1000; "(space)VOLTS"
```

The constant, 1.26, may be slightly incorrect for your circuit because of small variations in component values. To obtain a better value, apply a well-known voltage to the input (how about +5 volts?) and observe the measured frequency in KHz. Dividing the voltage by the frequency yields your constant.

Now that you have a voltmeter you can begin connecting it to transducers. You've been using a potentiometer to supply a sample voltage for testing, but how about using it as a transducer? If the element in the potentiometer has a linear taper, then you can use it without any modification or addition to measure angle. (The linear taper is specified as such when you buy it. If in doubt, rotate the potentiometer halfway through its range and see if the voltage output is half the voltage across the potentiometer. If it is, the potentiometer is linear.) Want to know not only if that door opened, but also how far it opened? Arrange to have the opening door turn the pot shaft. To calibrate it, determine the maximum angle through which it can rotate. This should be in the 270–330 degree range; you can measure it with a pointed knob and a protractor. Call this angle A. If V is the voltage supply to the pot and B is the angle you wish to measure, then the output voltage,

Vo will be $Vo = V*B/A$, or $B = A*Vo/V$. Said another way, the unknown angle in degrees will be A/V times the voltage you measure with the voltmeter. If you want to measure small distances, you can use a sliding pot such as the volume and tone controls on many radios and stereos. If D is the total distance the pot can travel and B this time is the unknown distance, then $B = D*Vo/V$. You can hook gear drives, levers, and other mechanical linkages to potentiometers to measure a wide range of angles and distances. You can even use rotating pots to measure distance, and sliding pots to measure angles if you can do the necessary trigonometry. Indirectly, sliding pots can measure other quantities, too. For instance, put a flat disk of some kind atop a coil spring or springs. Attach the slider of a pot to the edge of the plate. Now, when a weight is placed on the disk the coil spring will compress, and the slider attached to the edge of the disk will come down. You've just made a scale. Use known weights to calibrate it. A pound of sugar and five pounds of flour, for example, work quite well.

There is another way to build a scale that may not be as accurate as the one you just built, although conceptually it isn't much different. (It's still a resistive voltage divider.) This scale is a lot more fascinating, though, and it opens the way to designing a variety of pressure/weight sensitive transducers. It makes use of the conductive foam that packages CMOS circuits to prevent electrical damage from static charges. If you have a few bits of this foam around, you can use them. If you don't, Radio Shack sells a $5'' \times 5''$ sheet for about a dollar. It has the interesting property of simultaneously acting like a spring and changing its resistance when you compress it. You can construct this very simple scale as Figure 10.6 shows. The flat metal plates can be of almost any material. Copper works best because you can shape and cut it easily and solder it, too. Make sure that whatever you use is clean so that it will make good contact with the foam—use just enough pressure on the top plate to keep it in good contact. (This material may consistently spring back to its original thickness or it may take a set after repeated or prolonged use. The resistance change may be non linear, or the resistance may vary across the sheet; it may also vary from sheet to sheet). If you need more resistance for your application, stack two or more pieces atop each other. If you want less resistance, use a bigger piece or wire several pieces in parallel. As with the previous scale, you can calibrate it using known weights.

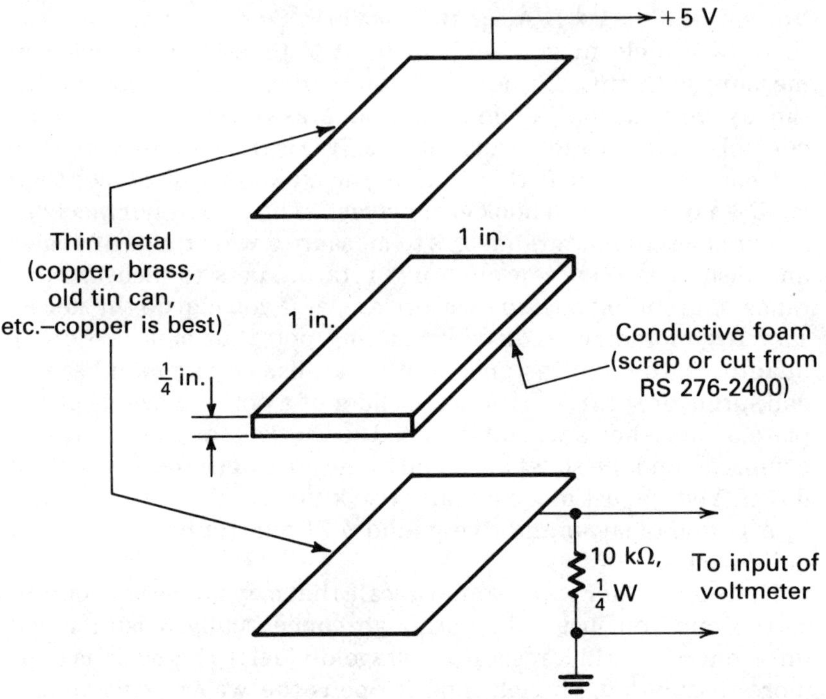

FIGURE 10.6 Conductive Foam Scale

Here are some other ideas you might try for different applications. Use conductive glue to hold the plates or electrodes in place. Bend a piece with electrodes attached to the ends rather than to top and bottom; this might work as a strain gauge or angle indicator. Place the scale in a closed container with a balloon and apply air pressure to the balloon; that should squeeze the scale, resulting in a change in voltage with pressure. Mount the scale between two fixed blocks of wood. If you then insert something between one of the blocks of wood and one of the metal plates, the voltage will be proportional to the thickness of the object, an electronic caliper.

There is another cheap resistive device that will let you measure the intensity of light. The cadmium-sulfide (CdS) photocell changes from a very high resistance, possibly several megohms, in the dark to as low as 100 ohms in bright light. The change is nonlinear, so a single proportionality constant won't work to calibrate it. You'll have to build a table of values into your program and interpolate between data points to get reasonably accurate measurements. You'll find the circuit diagram for the light meter in Figure 10.7. You'll note that it's identical to the one for the foam scale except that the resistor may have different values. Resistors in the 1K–10K range will work best for detecting variations in bright light, whereas larger resistors will work better for dim light. This is because the output voltage is Vout = 5∗R/(R+Rphoto), and Rphoto (the resistance of the photocell) is relatively small in bright light and large in dim light. When it comes to calibrating this device you'll either need another calibrated light meter or you'll have to work from the nominal specification of the cell, which is 1.7K at 1 footcandle (ftc) plus or minus 40%. Let's assume you don't have a light meter; if you did, you wouldn't be building one. If you let the resistor be 2.2K, then the output voltage at 1 ftc should be 2.82 volts. In a darkened room, move a 60–100 watt bulb toward and then away from the cell until you get this reading on your voltmeter. The reading at this point represents an incident light reading of 1 footcandle. If you double the distance of the bulb from the photoresistor, the light intensity will be one quarter as great; if you halve the distance, the intensity is four times as great. This is called the square law relationship.

$$I = k/D**2 \qquad I = \text{intensity}$$
$$k = \text{proportionality constant}$$
$$D = \text{distance}$$

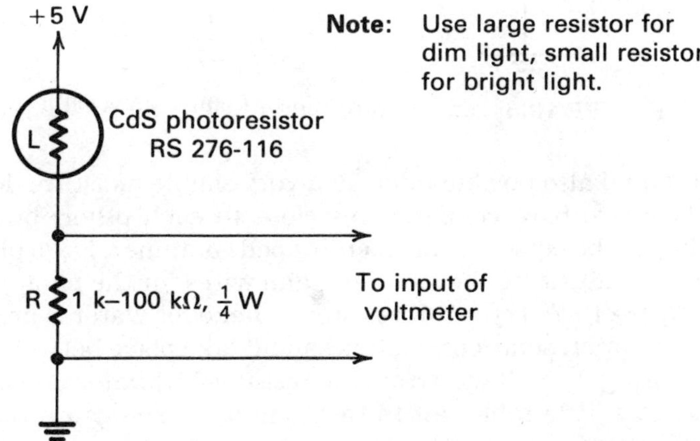

FIGURE 10.7 Light Meter Based on Cadmium-Sulfide Photocell

If you happen to have a thermistor, a resistor that changes value with temperature, you can hook it up exactly as you did the photocell and obtain a digital thermometer. Try varying the resistor to achieve the sensitivity you want and then calibrate it using a mixture of ice and water for 0 degrees centigrade and boiling water for 100 degrees centigrade. Use paint, glue, or some other substance to insulate the leads so they will not short out in the water.

If you can't get a thermistor, Radio Shack again has an inexpensive chip, the LM334, which you can use in its stead. Figure 10.8 shows that, except for the additional resistor, this circuit is the same as for the photocell. This particular circuit has a sensitivity of about 0.01 volts per degree centigrade, which means its output will vary only 1.0 volt between 0 and 100 degrees. It has yielded an output of 2.9 volts at 24 degrees.

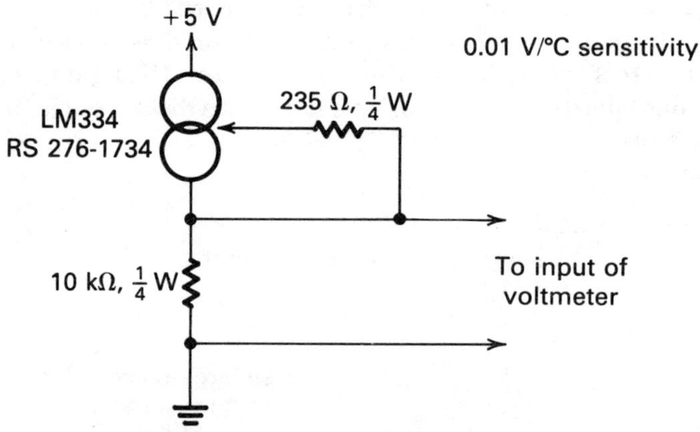

FIGURE 10.8 Temperature Sensor Using LM334 Chip

You should also be able to make a very simple moisture detector. Place two bare copper wires close to each other, but not touching, in the bottom of an open-topped container, like a plastic bottle cap. Substitute the container and wires for the photoresistor in Figure 10.7. Try a 10K resistor. Whenever water is present in the container, some conduction should take place between the wires, raising the voltage across the resistor. If this doesn't work, try adding a little table salt to the container. You can probably think of other ways to make a moisture sensitive resistor (drying out a small sponge previously soaked in a saltwater solution and attaching a wire to each end with an alligator clip), and you may even be able to buy one locally.

You'll undoubtedly run across other sensors that may be of use to you. As long as they put out DC voltage or a periodic signal you should be able to use either the counter or voltmeter to read them. If one of your measurements exceeds some specified threshold, you'll also be able to sound audible alarms using the tone generator that appeared earlier in this book. You can combine analog sensor measurements with discrete inputs via the KBD lines and create algorithms like ''If the door opens and the temperature is less than 15 degrees, then print a warning and sound an alarm.'' To put the machine codes for both the counter and tone generator in memory, you'll have to stack one above the other. That is, if the counter program extends to location 16538, then start the tone generator at 16539 and use USR 16539 to call it. All of the programs in this book use relocatable codes, so it really doesn't matter where in memory you put them.

Chapter 11

Affecting the Outside World

It is true that a sonic alarm or flashing video display affects the outside world to the extent that you, the system operator, will take some action upon perceiving it. What you'd really like to do, however, is cause the computer to directly manipulate something, some device, external to itself. Can the TS 1000 do this? Once again, the answer is yes, but only with some additional help.

The accepted method of helping the computer is to provide it with an output port. This is an input port in reverse. It takes information that exists only momentarily on the data bus within the computer and presents it to the outside world, generally as the presence or absence of a voltage or as the opening or closing of a switch. You can buy 3-byte (24-bit) programmable I/O ports for the TS 1000 for $70 to $80, or you can build one with four or five chips. These ports are a good value for $3 or $4 a bit, and they're convenient to install, as they clip onto the expansion connector at the rear of the computer. But you may have neither the money nor the skill and time for these approaches and, besides, you only need a couple of discrete outputs.

Fortunately, there are circuits that will provide discrete outputs at about the same cost per bit and do not require a great deal of skill for you to assemble. They're a little less convenient to use from a programming standpoint, and because they require several parts per bit they become cumbersome if you need more than four or five outputs.

To do this project, you need an amplifier capable of making audible tones when connected to the MIC (microphone) output of the computer. The Radio Shack unit suggested in the ''Tools and Other Unsightly Paraphernalia'' chapter works well. Use one of the jumper cables provided with the TS 1000 to connect the MIC output to the amplifier input. Attach two wires to an 1/8-inch miniature phone plug, and insert the plug into the EXTERNAL SPEAKER output of your amplifier. Make sure to note which wire is the ground and which is the signal lead. These leads are the input to the discrete port.

Generally speaking, if you wish your computer to actually move something, you need a transducer of some sort to convert the electrical signal from the output port into physical motion. Such devices are often called actuators. You won't learn how to build any, since they are frequently peculiar to the specific job at hand. However, closing a switch in series with a motor causes the motor to turn on. By means of levers, gears, pulleys, etc., you can make the rotary motion of the shaft do almost any job you want. A garage door opener is nothing more than a motor attached to a sprocket and chain, pulley and cable, or gear and rack. Other types of actuators are available from a variety of sources: solenoids for small linear movement; solenoid valves for controlling gas or fluid flow, linear motors for intermediate range motion, DC motors for variable-speed movement (this would involve outputting stepwise or continuously variable DC—*Direct Current*— voltages), stepping motors and relays of discrete angular motion. You've probably thought of still others already. The rest of this chapter will, therefore, concentrate on the output ports themselves. A block diagram of the systems you'll build is shown in Figure 11.1. Note also that, with one exception, you've already learned the programs you need to activate the ports.

You can use the first circuit only if you need a single port and can run your program exclusively in FAST mode. You activate the output by placing the computer in SLOW mode. This is not quite as restrictive as it may sound. For instance, when you go to bed at night you might run a program that simultaneously monitors trip switches on your windows and doors in case of a burglary attempt and on a heat detector in case of fire. You might even want it to awaken you for breakfast. There's no sense running the TV when there's no one around to see it, so FAST mode is fine. To warn or wake you, you'll want a loud Klaxon to sound. The program line that trips the alarm will be something like

xxxx IF (#1 occured OR #2 occurred, etc.) THEN SLOW

The circuit diagram appears in Figure 11.2. The load is shown as R, which was 100 ohms for test purposes but could be any DC load requiring up to 6 amps (amperes) at 400 volts if your power supply can handle it. Put a heat sink on the SCR if you are planning to draw more than 100 ma (milliamps) or so; a $5'' \times 5''$ piece of aluminum will do. If you plan to switch AC (*Alternating Current*) use a reed relay for R and use its contacts to switch the AC load. (In principle, you can use a triac to directly switch the AC load, but connecting AC lines directly to electronics endangers the equipment and can knock you for a loop if you're not careful.) Radio Shack catalog numbers 275-240, 275-243, and 275-246 will all work with a 5-volt supply. *Do not* use the computer's 5-volt supply unless you have added the preregulator: these relays draw almost 100 ma. The amplified synch pulses coming from the computer when it enters SLOW mode trigger the SCR, causing it to conduct current. It will continue to conduct until the current is reduced to zero; hence, the normally closed reset switch. If, when you push the RESET button, the computer is back in FAST mode, the current will remain off after releasing the reset switch and will stay off until you again enter SLOW mode.

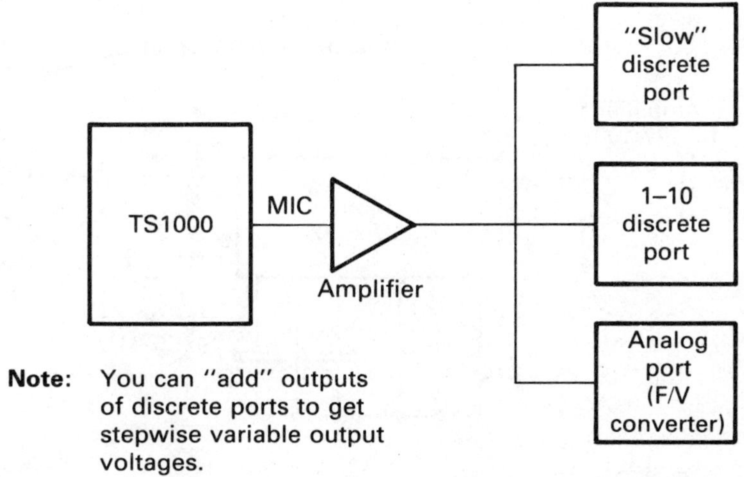

FIGURE 11.1 Block Diagram of Output Systems

Now, suppose you want to drive a small motor for a short, fixed period of time (as long as it takes to open the damper on your wood burning stove to warm the house before you get up in the morning), sound an alarm until you manually turn it off (a wake-up

alarm), and turn on a light that will go off when you leave the
house. Can the TS 1000 do this? Sure!

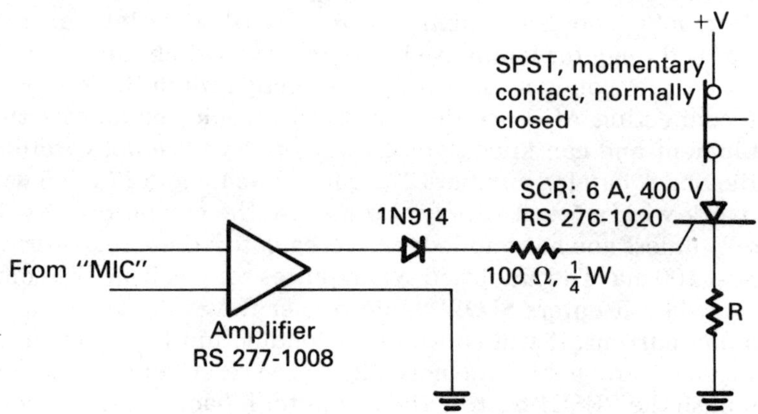

FIGURE 11.2 Single-Bit FAST-Only Port

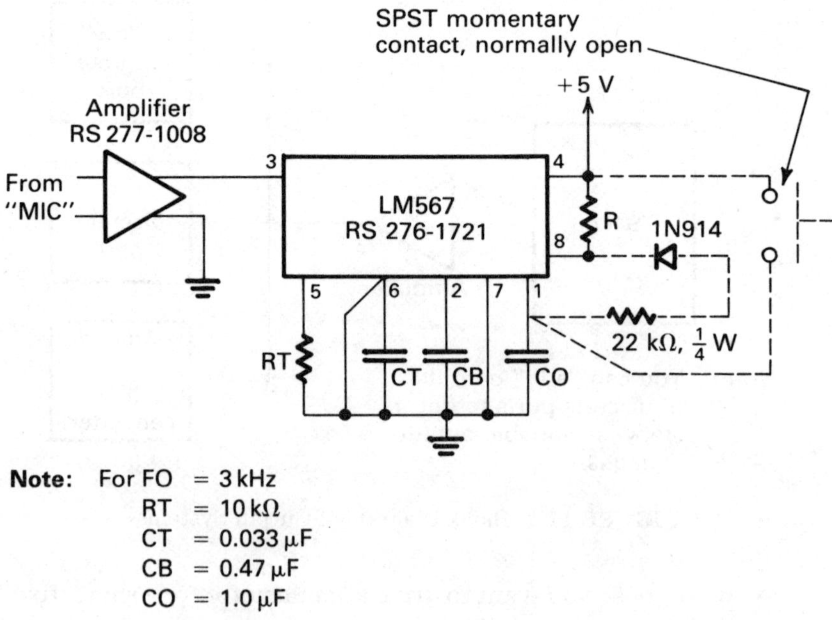

Note: For FO = 3 kHz
            RT = 10 kΩ
            CT = 0.033 µF
            CB = 0.47 µF
            CO = 1.0 µF

FIGURE 11.3 Single-Bit Tone Decoder Port

You must build three circuits similar to the one in Figure 11.3. You need only one amplifier to drive all three circuits, however. The IC (*Integrated Circuit*) is an LM567 phase-locked-loop tone decoder. The output of the chip at pin 8 will remain at or near the power supply voltage until it receives a tone of a particular frequency, and then the output will go to a few tenths of a volt above ground. The external components determine which frequency or band of frequencies will trigger the device. RT and CT determine the center frequency. Pick a value for RT between 2K and 20K, choose the center frequency, FO (in Hz), and find CT from:

$$CT = \frac{1,100,000}{RT * FO} \text{ (microfarads)}$$

CB determines the bandwidth, or range of frequencies that will be accepted by means of this formula

$$CB = \frac{n}{FO} \text{ (microfarads)}$$

where n ranges from 1300 (for a 14% bandwidth) to 62,000 (for a 2% bandwidth). Make CO = 2 * CB. The resistor, R, may be any load, in particular a 5-volt reed relay, which does not draw more than 100 ma at 5 volts. For the values in Figure 11.3, the center frequency is 3 KHz, and the bandwidth is 420 Hz (14% of 3000). You can use narrower bandwidths, but component variations will cause the center frequency to be somewhat different from the calculated value. For this reason, the bandwidth has been left wide enough to encompass the actual frequency. Should your decoder not work with a 3-KHz tone try varying the frequency in 100-Hz steps, either side of 3 KHz, for about 400 Hz.

If your load is a relay, then its coil will be activated and its contacts picked as long as a 3-KHz tone is present at the input to the circuit. The tone should also rise above a 2-volt peak. Note in particualr that the TV synch pulses are not within the bandwidth, so your program can run in SLOW mode without triggering an output. Now, if you use a relay that's normally open, you can use the contacts to switch the AC line onto a small motor, and the motor will continue to run for as long as the tone persists. If you use the tone burst program (Program 7.4 from ''Sound Advice''), you can generate about 20 seconds of 3-KHz tone. If you need greater durations you will have to loop on the USR routine.

The program you must construct will be a combination of the 24-hour clock and tone bursts, and it will be configured something like the one that follows. (No, you won't get the completed program because it probably wouldn't fit your exact requirements anyway.)

```
     10   REM (tone burst routine) + (timing routine)
aaa–bbb  (clock program)
ccc      IF (time of day) = (open damper time) THEN GOSUB rrr
qqq      GOTO (reentry to clock program)
rrr      LET N = (frequency number)
sss      LET HILO = (length number)
ttt–uuu  (statements 60–85 of tone burst program)
vvv      RAND USR 16514
www      RETURN
```

So, you've opened the damper and, some time later, with the old homestead toasty warm, you'd like your computer to sound a wake-up alarm until you manually turn it off. You begin by building a second tone decoder and you modify it by adding the parts shown as dotted lines in Figure 11.3. The diode and resistor cause the output to latch for any tone of the proper frequency, provided it's longer than 5–10 msec (milliseconds). That is, the output will stay low, even after the tone has disappeared, until you push the reset switch.

Design this decoder using the equations for 2-KHz operation. (Hint: changing CT to 0.05 microfarads will probably work.) Substitute a 5-volt buzzer or Sonalert for R, and use any convenient momentary contact switch for the reset; a keyboard switch will work nicely. Remember that the buzzer must draw less than 100 ma or you will have to use a relay to activate it.

Add another test to the program.

```
ddd  IF (time of day) = (get-up time) THEN GOSUB mmm
```

The subroutine at mmm will be exactly the same as the one at rrr, except that the frequency and length numbers will correspond to 2 KHz and 50 msec.

By the time you've finished your morning rituals in the bathroom, you'll want a light on in the hall and you'd like the light to continue burning until you leave the house, at which time it should shut off automatically. Build another decoder with the latching components, but set its center frequency for 4.5 KHz.

Use a relay for R, and have its contacts switch on the light. Add another test

     eee IF (time of day) = (time for light) THEN GOSUB kkk

and another subroutine like the last, but with numbers for 4.5 KHz and 50 msec. This time, mount the reset switch so that it is closed when the door through which you leave the house is opened. The light will go out as you leave.

This example may be of a slightly contrived nature, but you can readily see how the tone decoders make cheap and flexible single-bit output ports. If you think about it for a minute, you'll see that it also illustrates one of the big problems with getting computers to do things; namely, the necessity of running wires all over the place. To make the example work you'd need to run two wires from the computer to the stove, three to the alarm, two to the light, and another two from the door to the decoder. This is not a function of the computer you are using. The same wires would be required if you had an IBM personal computer with every option available. Those wires must be there to carry signals between the computer and the actuators and transducers. Only in the case of the light could you reduce the number of conductors by changing your strategy slightly. Instead of using a door switch to reset the light decoder you could build another decoder to reset the first (the relay of the second would momentarily short pin 8 of the first decoder to pin 4) soon after the hour you generally leave the house. Figure 11.4 is a block diagram of your morning control system.

There are also ways in which you can get different voltage levels out of your computer. Figure 11.5 shows three methods of using voltage dividers and decoders to achieve different output voltages. The first two do not require load resistors or relays in the decoders themselves. Naturally, the more decoders you have, the more voltages you can obtain, but because you can't activate more than one decoder at a time, you're stuck with $N+1$ voltages given N decoders. A limited number of step-motor speed control could be achieved in a similar fashion.

If you need more than two or three voltages or if you must have a continuously variable voltage, then forget about these methods and look at the applications note that comes with the 9400 V-F/F-V converter. It describes a circuit that will convert an input signal of a given frequency into a DC output that's linearly proportional to the input frequency. Build the circuit and attach it to the amplifier output just as you did the tone decoders. You can arrange the scaling with the external components to get the desired voltage for the range of frequencies the TS 1000 can produce.

While all of these approaches work nicely, they do require an amplifier, and the range of frequencies you can generate is somewhat limited, both effects caused by the bandpass filter at the MIC output of the computer. You can get around this requirement by going into the computer and tapping off the signal before it is filtered. But there is a problem: the vertical synch generator you toggle to generate your tones. Whenever it goes high, hardware within the TS 1000 begins to automatically generate horizontal synch pulses, which are superimposed on any signal you may try to form. These pulses occur every 64 μsec (microseconds) and are 5 μsec wide. They are not present at the MIC output because that bandpass filter removes them.

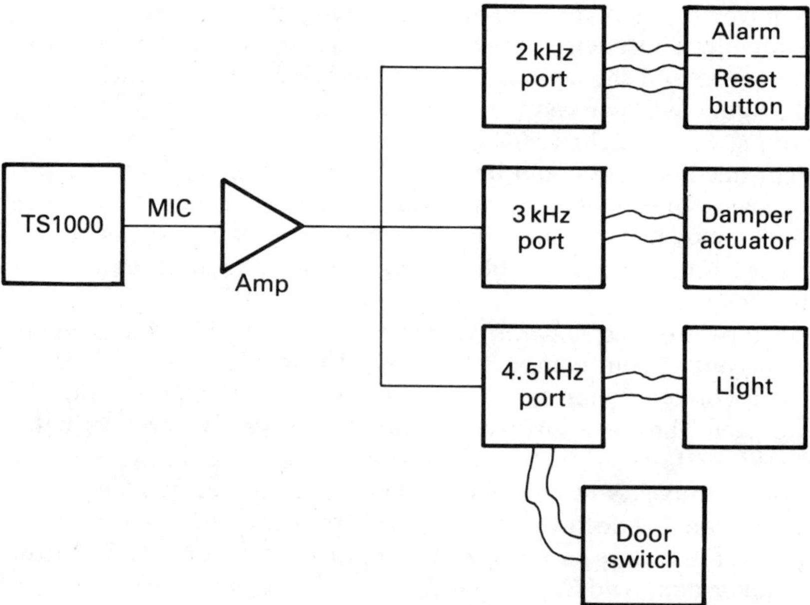

FIGURE 11.4 Block Diagram of a Hypothetical Morning Control System

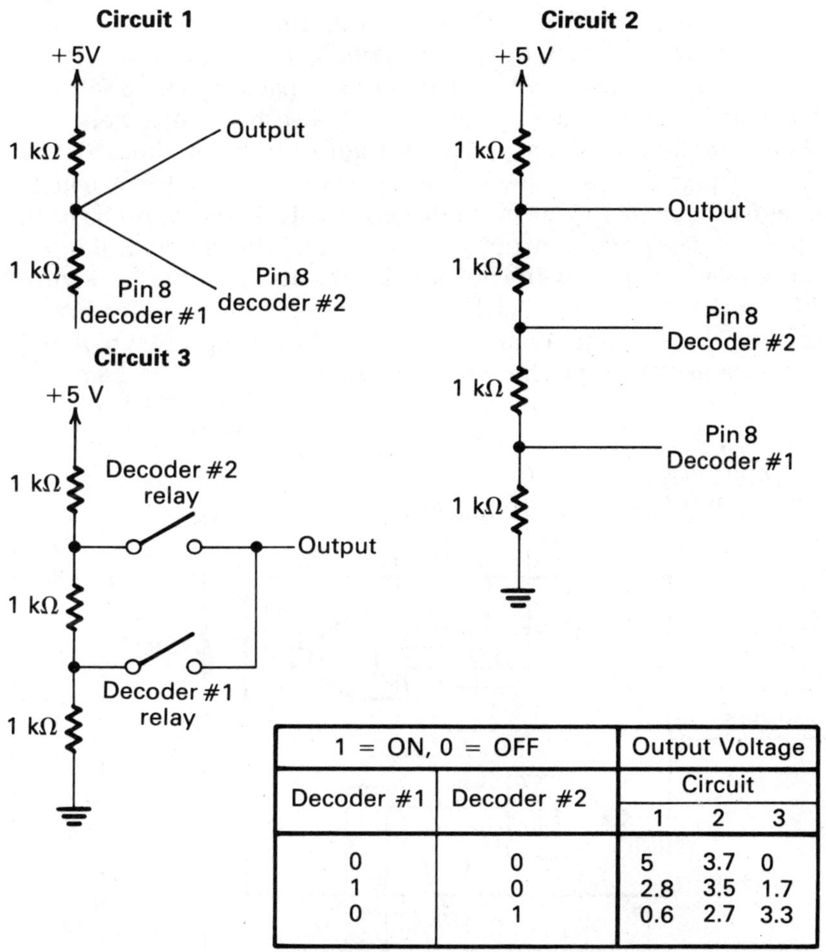

1 = ON, 0 = OFF		Output Voltage		
		Circuit		
Decoder #1	Decoder #2	1	2	3
0	0	5	3.7	0
1	0	2.8	3.5	1.7
0	1	0.6	2.7	3.3

FIGURE 11.5 Discrete Voltage Outputs

Since you'll find it impossible to create any signals less than five μsec long by toggling the vertical synch generator, you'll build a circuit that ignores signals less than 5 μsec in duration. The circuit will also make a handy output buffer. The schematic for it appears in Figure 11.6. Incidentally, if you have a MicroAce or ZX80, you don't need this circuit, although it is a good idea to use it as a buffer in modified form by omitting the diode, resistor, and capacitor and tying pin 6 directly to pins 9, 10, 12, and 13. On the MicroAce you'll attach the input to pin 5 of U13, pin 1 of U14, or the end of R33 that's not attached to a capacitor. The ZX80 will have similar attachment points. You'll also be able to generate slightly higher frequency signals than owners of the TS1000/ZX81. The circuit, as shown, extends any positive pulse by a little over 5 μsec so that if the input drops for only 5 μsec it will be back up before the circuit output goes down, and the downward input pulse won't appear at the output. The location of D9 is in Figure 13.2 of the "Potpourri" chapter. If you've built the video inverter, use the alternate connection at pin 4 of the 74LS02 to avoid unnecessarily loading the synch generator.

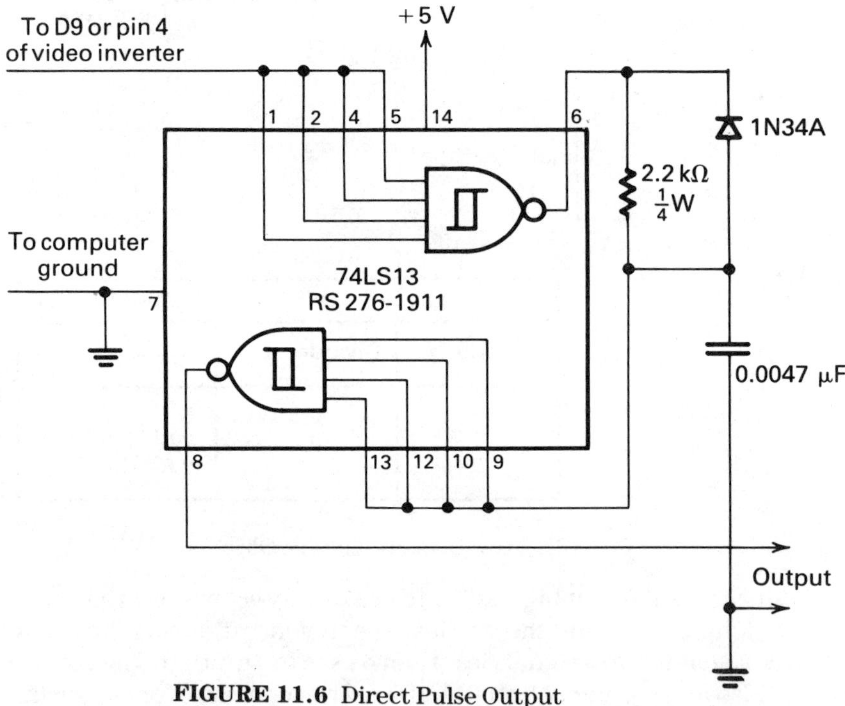

FIGURE 11.6 Direct Pulse Output

A big advantage of this circuit is that you can directly generate 0- to 5-volt pulses with accurately timed durations of 19 μsec to as long as you wish, provided you don't enter SLOW mode. You can even develop machine code routines that will serially transmit data or instructions from your computer to peripheral devices like printers and terminals, but be careful what you connect your computer to because the 74LS13 will only handle 8 ma at low-level output and only 0.4 ma at high-level. Follow the 74LS13 with a driver transistor if you need more current.

The tone burst program will still work, but the signal will be unsymmetrical because you've lengthened the positive portion at the expense of the zero portion. You can almost regain symmetry by changing line 60 to **POKE** 16523,N+1 and line 70 to **POKE** 16529,N+1. This should work satisfactorily with the tone decoders. You can get almost perfect symmetry by adding two more NOPs right after the one in the machine code, but it will make the period of your tone 2.48 μsec longer. (In calculating N and length, 87 will replace the 79.) If you try this, remember to alter the JR C,−16 to JR C,−18 so the program jumps back to the right spot.

With this circuit you can use the TS 1000 as an audio pulse generator, which is a handy instrument if you're into electronics. Program 11.1 will let you generate square waves from 1 Hz to 21.5 KHz, and up to 31 KHz for pulses with a 71% duty cycle. The mnemonics for the machine code routine in this program follow.

Starting Location	Code	Mnemonic
16514	17,255,255	LD DE,−1
16517	62,127	LD A, 127
16519	219,254	IN A,254
16521	31	RRA
16522	208	RET NC
16523	33,0,0	LD HL,0
16526	25	ADD HL,DE
16527	56,−3	JR C,−3
16529	211,255	OUT 255,A
16531	33,0,0	LD HL,0
16534	25	ADD HL,DE
16535	56,−3	JR C,−3
16537	24,−22	JR,−22

This routine is nothing more than two HL delays. An IN A (reset synch) precedes the first and an OUT A (set synch) precedes the second. Control then jumps back to 16517, the BREAK key is tested, and the loops repeat. The BASIC portion of the program inputs the parameters that determine the set time (N1) and the reset time (N2), converts them to high-order and low-order bytes, and **POKE**s them into the appropriate locations. The pulse width and period in μsec are given by

$$PW = 23 + 7.077 * N2$$
$$T = 32.3 + 7.077 * (N1 + N2)$$
$$N1 = N2 + 2 \text{ for square waves}$$

When the program is pulsing, depressing the BREAK key will halt it. You can, of course, enter the pulse width and period directly into the program and have the computer do the calculation of N1 and N2 for you. You can try that as an exercise on your own.

PROGRAM 11.1 Pulse Generator

```
10 REM (25 characters)
20 PRINT "INPUT N1"
30 INPUT N1
40 PRINT "INPUT N2"
50 INPUT N2
60 POKE 16525,INT (N1/256)
70 POKE 16524,N1 – 256 * INT (N1/256)
80 POKE 16533,INT (N2/256)
90 POKE 16532,N2 – 256 * INT (N2/256)
100 FAST
110 RAND USR 16514
```

Sequentially **POKE** in the following codes beginning at location 16514.

17, 255, 255, 62, 127, 219, 254, 31, 208, 33, 0, 0, 25, 56, 253, 211, 255, 33, 0, 0, 25, 56, 253, 24, 234

Many other pulse generation schemes are also possible. You could use the tone generator in Program 7.2. The BASIC would have to be changed so that the values of N1 and N2 were **POKE**d into locations 16522 and 16528. The parameters would then be given by

$$PW = 15 + 4 * N1$$
$$T = 17.6 + 4 * (N1 + N2)$$
$$N1 = N2 + 3 \text{ for square waves}$$

This routine has the advantage of being able to generate square waves up to 26.3 KHz and pulses up to 39 KHz with a 74% duty cycle. The frequencies generated are also closer together as N1 or N2 increment. Unfortunately, the tone generator only descends in frequency to 487 Hz.

For special purposes, you can generate much shorter pulses. For instance

```
OUT 255,A
IN A,254
JR,–6
```

will generate a 3-μsec negative-going pulse every 10.5 μsec, but be careful here. This is as fast as you can get the TS 1000 to toggle; the narrow pulse is a result of both the coding and the pulse length extender in Figure 11.6. The pulse generated within the computer appears as a solid line in Figure 11.7 (superimposed synch pulses ignored). The pulse at the output of the buffer, after extension, is the dashed line. What you've really done is generate a 2.5-μsec positive-going pulse, extend the pulse reset time by 5 μsec, and then repeat the process 3 μsec later. This is as short a pulse as you can generate without changing the extender. You can conveniently change the period, however, by putting a delay between OUT A and IN A. If you put the delay between IN A and JR instead, then you can generate 7.5-μsec positive-going pulses with a period determined by the delay. The JR, in either case, should still be back to OUT A, so remember to change the –6. Note that no provision is made to check the BREAK key.

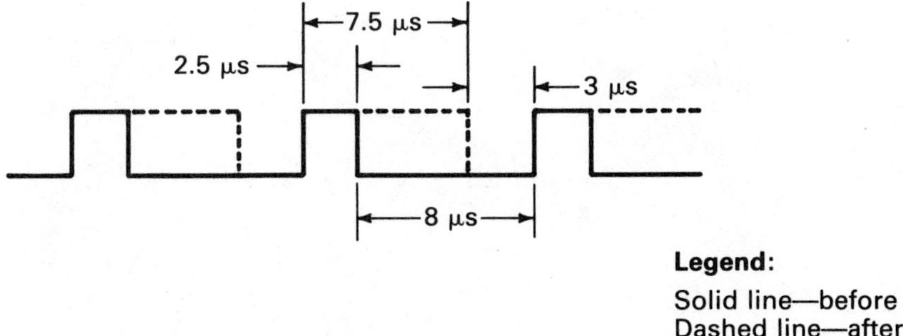

FIGURE 11.7 Pulse Width before and after Pulse Extender Circuit

Chapter 12

Saving Grade

It's apparent from the letters to the editor in many ZX81/TS 1000-oriented magazines that a number of users have experienced some difficulties in **SAVE**ing and **LOAD**ing tapes. Although this seems to be the exception rather than the rule, you may find this section on recording tips and add-ons helpful.

If you haven't yet purchased a recorder, there are some particular features you should look for.

1. The recorder should run on AC (*Alternating Current*) power or four flashlight batteries. Minirecorders using only two cells may not provide a high enough output voltage, and AC power saves on battery costs.

2. Try to find a recorder with an ALC (Automatic Level Control) or AVC (Automatic Volume Control) that can be switched out. ALCs tend to run the gain full up during the quiet period that is recorded on the tape preceding the **SAVE**d program. They then overdrive and distort the beginning of the program before their circuitry can compensate for the changed signal level.

3. Fast forward and reverse cuing (the ability to hear the tape while in fast forward or reverse) is very useful when searching for a program.

You can use a home hi-fi stereo recorder, but keep in mind these potential problem areas.

They are usually not portable enough to be placed adjacent to the computer. They are also usually hooked into other hi-fi equipment. The need for longer cables (stretched across the

room and possibly tripped over) and the multiple interconnec-
tions will increase the chances for hum and increase the radio
frequency noise (the stuff that causes wiggles on your TV and
whistles on the FM) put out by the computer.
The output of the tape deck itself will probably not be sufficient to drive your computer; you'll need an amplifier. If you use
the attached hi-fi amplifier, you'll have to ensure that the
output signal doesn't exceed about 5 volts peak, or you'll risk
damaging your computer.

Assuming you now have your recorder set up and ready to go,
follow this procedure.

1. Make sure the head is clean and not worn. A very dirty head
will have a buildup of the reddish-brown material from the
tape on it. You can clean it off using a cotton swab and alcohol
or freon, or you can buy a special head cleaning cassette. Don't
overuse a head cleaning cassette, as it's slightly abrasive and
can lead to shortened head life. A badly worn head will show
ragged metal edges surrounding the head gap. A good head
should have a crisp-edged slit down the middle of it. If the head
is shot, replacement is the only answer.
2. If you have a tone control, set it to full treble.
3. If you're using a stereo recorder, don't use the monaural
("mono") mode. Instead, short one input with a shorting plug,
and record on the other channel.
4. Load the **SAVE**-like signal generator from the "Sound Advice" chapter and start it running. If you have a record signal
level meter on your recorder, set the controls so that the needle
rests just barely on the low side of the red (0 db—decibles). Play
the signal back. On most portable recorders you will hear an
uncomfortably—but not painfully—loud, raucous buzz-tone.
You are now ready to **SAVE** a program.
5. Type in a few program lines. They can be meaningless REM
statements or a program. Run the tape for several inches, and
place a voice header ("This is a test") on it. Do not attempt to
SAVE programs at the very ends of a tape, since this is where
most tape errors will occur. Type **SAVE** "TEST" but don't
ENTER. Make sure your MIC (microphone) lead is plugged into
both computer and recorder. Start the recorder and hit enter.
When **SAVE** is complete stop the recorder and rewind the tape.
Remove the EAR (earphone) lead so you can hear and play back

the tape. You should hear your voice header, a buzz, about 6 seconds of silence, the program sound, a buzz, and blank tape, in that order.

6. Rewind the tape and play it back to the end of your voice header. Plug the EAR cable into both recorder and computer. Type **NEW**, **ENTER**, **LOAD** ''TEST''. Start the recorder and hit **ENTER**. You'll see a pattern of some kind on the TV screen. It will change to a more random pattern at a point on the tape where the program is actually recorded. When the program has **LOAD**ed, the screen should go white with a 0/0 in the lower left corner. Press **ENTER**, and the program listing should appear.

In case these directions don't work quite right for you, here are a few troubleshooting hints.

Hum is a frequent problem. Usually you'll become immediately aware of hum problems when you hear the loud, 60-Hz sound that comes from your recorder on playback. If you're not sure, though, try this.

```
10 FAST
20 IF INKEY$ < > " " THEN GOTO 20
30 IF INKEY$ = " " THEN GOTO 30
```

Attach both cables to your recorder and start it recording. **RUN** the above program for several seconds. Hit any key to stop it. Rewind the tape, and play it back with the EAR plug removed. You should hear a buzz until you start the program and, afterward, just silence or a very faint hiss. If you hear anything else you have hum or noise problems. Hum will sound like a low-pitched tone; noise will be a loud hiss or possibly a crackling sound. First, make certain you've inserted both ends of the cables properly and fully in their receptacles, and then repeat the experiment. If hum is still present, reverse the AC (*Alternating Current*) power plug on the recorder (if possible). Try the same for the TV plug. Failing that, remove the EAR cable while recording and the MIC cable while playing back. If that works, and you don't like having to remove the cables, use an X-acto knife or razor blade to carefully remove about one inch of outer insulation from the EAR cable. Then, very carefully sever the exposed metal shield, but not the underlying insulation. Tape over the exposed spot, and your problem should be cured. If removing a cable didn't work, then run the recorder on battery power with the line cord unplugged. If you still have hum, borrow another recorder and try again. If the problem goes away, your recorder has internal problems. If it's still present, your computer has internal problems.

Another problem may be lack of sound on playback while the TV shows a characteristic pattern during **SAVE**. Check for the record level control turned all the way down, a cable unplugged or broken, an internal recorder problem, or an internal computer problem, in that order.

Here are some more symptoms: sound on playback; a TV pattern characteristic of a silent period during **LOAD**; a program that won't **LOAD** even though you've used the recorder successfully in the past to **SAVE** and **LOAD**. Check for the volume control set too low, the tone control set at full bass, an unplugged or broken cable, a dirty or worn head, a bad tape, and a recorder or computer breakdown, in that order. If you notice these symptoms in a recorder you haven't used before, verify that you have good cables, proper control settings, and a good tape. If you have no luck, try again with a borrowed recorder.

Possibly the programs you've written and **SAVE**d will **LOAD**, but other tapes won't. This may indicate that your recorder or the recorder used to make the other tapes has tape speed or head azimuth problems. If no other tapes will load, it's probably yours. If just one or two tapes won't load (especially if they're from the same other recorder), it's probably the other machine. You can check your recorder's speed against the other recorder's. Load the tone generator program from "Sound Advice," **POKE** 16522 and 16528 with 31 and run the program. Start your recorder, and record several minutes of the 3.717-KHz (kilohertz) tone you're producing. Load the frequency counter program, and measure the frequency on the tape you've just recorded. It should be well within 50 Hz (hertz) of what you recorded and should not vary more than that from sample to sample; otherwise, you've got intolerable speed fluctuations. Now play the tape back on the other recorder and measure the frequency. If it differs more than 50 Hz, then one or both of the machines has a speed problem. If you think your speed is okay and you simply *must* load that tape, you can try altering the head azimuth. To do so, you must loosen the screws holding the head and tilt the head relative to the direction of tape travel. It's a time-consuming process, since you don't know how much or in which direction you must tilt the head, and there's no guarantee of success without test equipment. You must also be able to put the head back in its original position when you're done. You should probably not attempt altering the head azimuth, unless there is no other option.

If you're using a stereo tape deck or minirecorder and the output is too low, see the "Potpourri" chapter for a simple amplifier.

Last, if you object to plugging and unplugging cables every time you want to add a voice header, listen to play back, **SAVE** or **LOAD**, you may want to assemble a simple switch box. Here's one you can put together in a small coffee tin with the speaker facing up and so that the perforated plastic lid will let the sound through. (You can use any container you desire.) You'll find a schematic and sketch of the wiring in Figure 12.1. The single switch lets you choose whichever mode you need. You both speak into and hear from the speaker; the auxiliary input jack is optional. The box is a real convenience. Note also that the EAR cable shield does not run through the box, thereby minimizing chances of hum problems. The parts the box requires are inexpensive.

1	6-position, 2-pole rotary switch (Radio Shack catalog number 275-1386)
1	4-inch-round PM speaker, 8-ohm Philmore T531 (available at Radio Shack)
1	1/8-inch miniature phone jack (Radio Shack catalog number 274-296)
1	1 K-ohm, 1/2-watt resistor (available at Radio Shack)
1	100-ohm, 1/2-watt resistor (available at Radio Shack)

You can use the cables and plugs that came with your computer. If the 100-ohm resistor fails to provide the proper sensitivity (so that you don't have to change the volume on playback and record) then increase it to get higher volume and decrease it for lower.

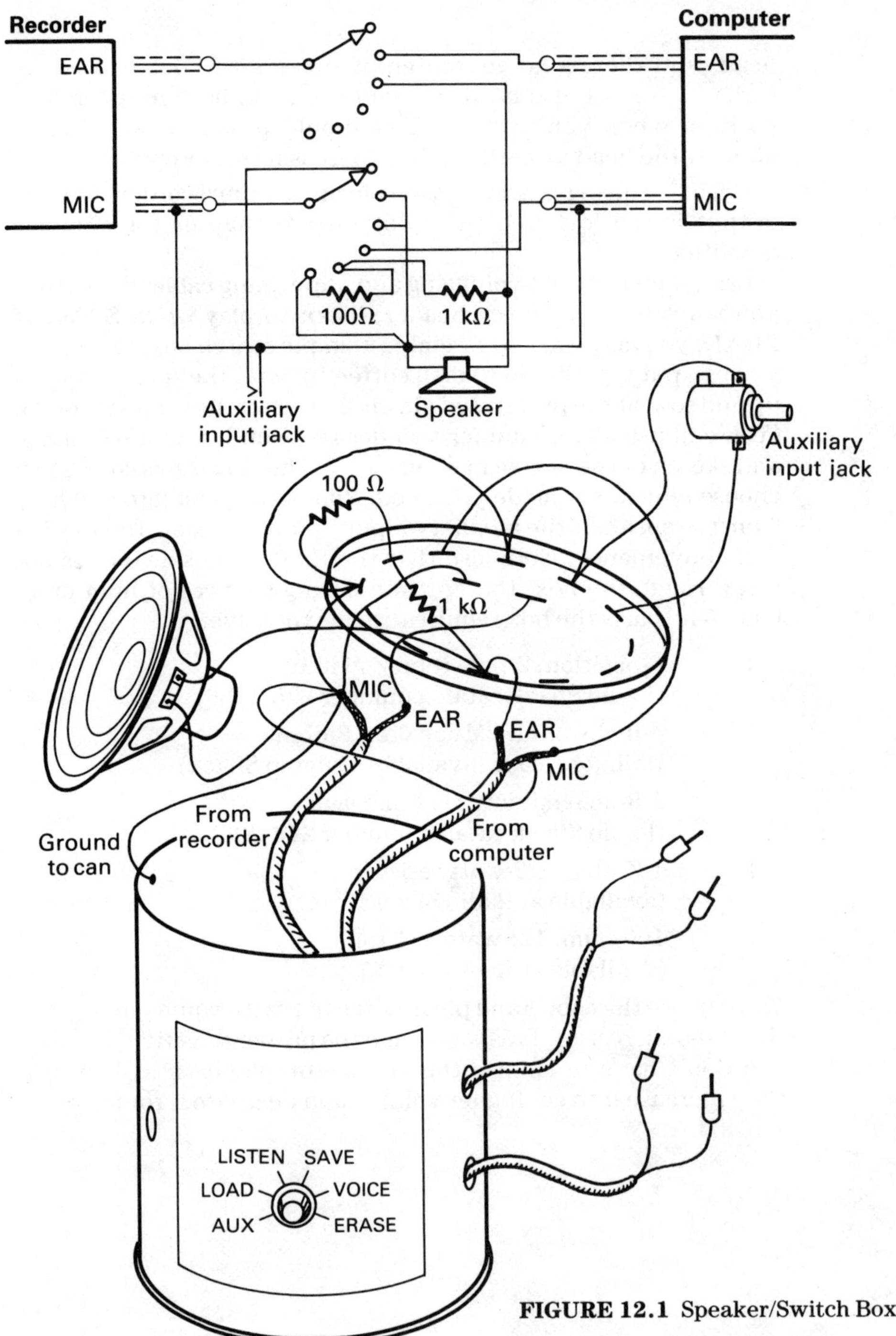

FIGURE 12.1 Speaker/Switch Box

Chapter 13

Potpourri

Besides all the projects you've already tried, there are a few circuits and programs that didn't seem to fit within the context of any of the other chapters. Here, then, is a chapter devoted to these misfits.

First, there's the simple tape amplifier you'll see in Figure 13.1. The first section of the LM3900 works as an open loop amplifier, which provides gain of 30–60 for input signals at about 0.1 volt peak-to-peak. The second section simply detects when the signal peaks out of the first exceed 2.5 volts and lights the LED (*Light-Emitting Diode*). Recorder volume and tone settings remain critical. Increase the recorder volume until the LED just flashes and then increase it a tad more. (Further increases will result in recorder/tape noise being amplified too much.) Using the amplifier does give slightly more uniform pulse tops because it drives them to saturation for high signal-to-noise ratio inputs. You can use the second half of the circuit by itself as an indicator of correct volume setting, whether you use the first half or not. Connect the recorder output to the 1K resistor, and delete the first half of the circuit.

You've probably heard that a video monitor will give a crisper display than a TV set. While this is generally true, most TV sets, especially the newer models in good repair, are quite capable of yielding all the resolution you need for a 32-character line. A video monitor will not, therefore, significantly improve the picture you get from your TS 1000, again assuming you've adjusted everything properly and made your connections correctly. Do not buy a monitor if this is your goal. If, on the other hand, you must have that green or amber screen, if you're contemplating buying a more advanced computer in the future, or if you already have a monitor and would like to use it instead of tying up the family TV for hours on end, then you can easily add a composite video output to your computer.

Open your computer and solder the shield of a short piece of coaxial cable to the ground as Figure 13.2 shows. Attach the center conductor to the junction of R31 and R32. Mount a jack of your choice in the side of the top half of the case opposite the normal connectors, and attach the other end of the coaxial cable to it. RCA, BNC, and TV F-type jacks are all appropriate. Make up a cable with a mating connector to run to your monitor. If your monitor already has a high-impedance input, that's all there is to it. Jack in and start typing. If it doesn't, then you'll have to modify it, because the TS 1000 will be loaded down by the 75-ohm inputs that are standard on most monitors. Fortunately, most monitors realize the 75 ohms through the simple expedient of shunting the input with a 75-ohm resistor. Both Zenith and Conrac use the shunt technique, and most other manufacturers may too, since the inherent input impedance of the first video amplifier will usually be reasonably high (high being defined as greater than a couple of kilohms). If in doubt, find a schematic of the monitor and check. All you need do is disconnect the low-value resistor (75 ohm or 120 ohm, or whatever the low resistance value may be). If you want to preserve the low Z input, install a switch to cut the 75-ohm resistor in or out of the circuit. With the modification complete, you can begin typing. The RF modulator remains in-circuit, so you have not lost the capability of using the TV.

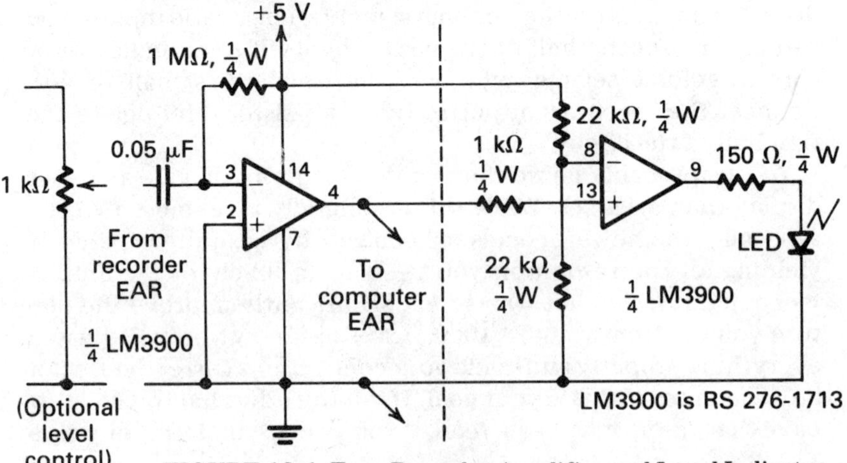

FIGURE 13.1 Tape Recorder Amplifier and Level Indicator

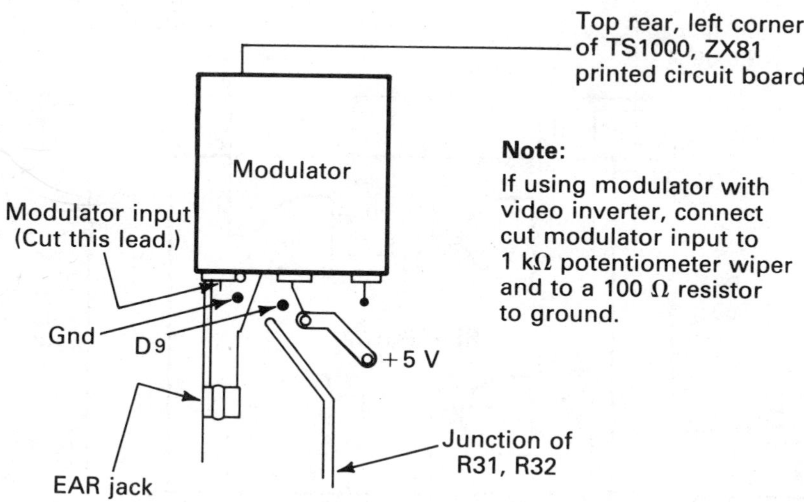

FIGURE 13.2 Video Attachment Points

You've doubtless noticed that while you can type the video inverse of any character under keyboard control, you cannot invert the whole screen. There is no easy modification for this as there is in the case of the MicroAce and ZX80. (The MicroAce has three holes near the center of the board labeled A, B, and C, with A connected to B and C by white lines. Cut the existing copper land between A and B, and install a small switch to connect A to B or A to C for normal or inverse video. A similar method should be possible with the ZX80.) Instead, you must build a circuit that separates the synch pulses and the video, inverts the video, and then recombines the two. This circuit appears in Figure 13.3; the connection points are in Figure 13.2. The wiper of the 1K pot must see a low impedance, about 100 ohms. If it's used with a 75-ohm monitor, it will work fine. If your monitor is high-impedance, or if you are going to continue using the RF modulator and the TV, then ground the wiper through a 100-ohm resistor. You can mount the circuit either inside or outside the computer. If you're using yours without its case glue a piece of foam to the bottom of the perfboard on which you build the circuit, and glue that to the top of the modulator. When using the device, set the pot as near as possible to its center position. You may have to readjust the pot or the contrast and brightness controls of the TV/monitor when switching from normal to inverse.

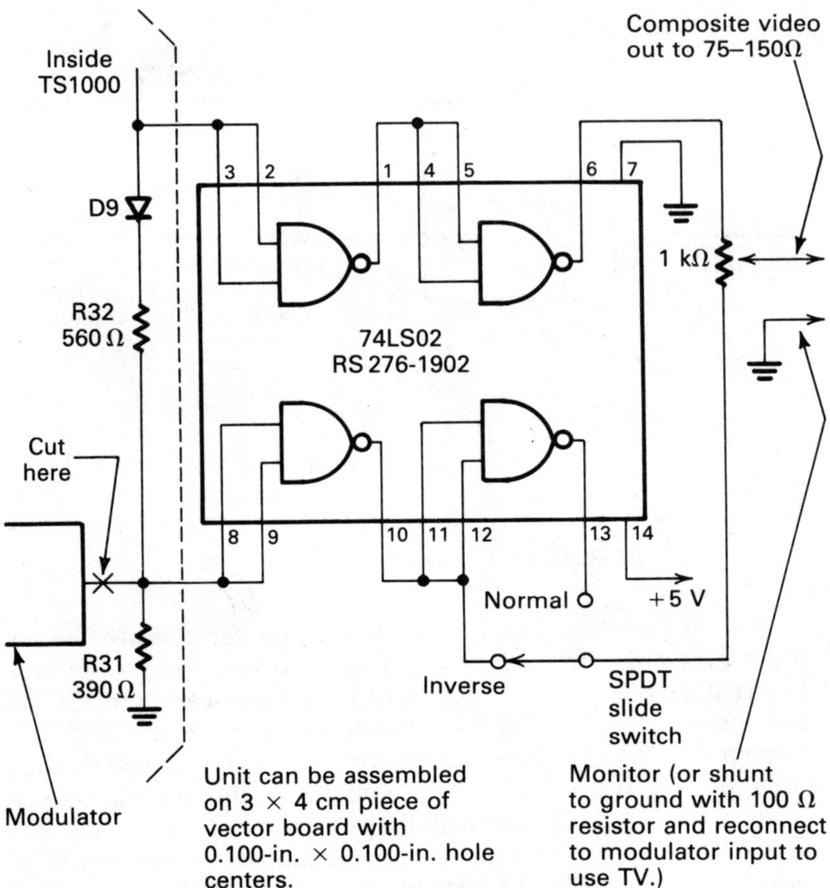

FIGURE 13.3 Video Inverter

Last, there are three very simple programs that may be of some
use when you adjust your TV. They are Programs 13.1A, 13.1B,
and 13.1C. The first, Program 13.1A, produces an equi-spaced
array of black dots ten wide by seven high on a white field, each
dot 4 × 4 points in size. If the dots are not equispaced (measure the
distance between them with a rule to check), you probably need
to tinker with the linearity control. You can also use the program
for horizontal centering, width, and height adjustments. It's not
very useful for vertical centering, since it won't print the last two
rows of dots or print either vertically or horizontally into the

overscan regions of the screen. These are limitations of the firmware in the TS 1000, and you can circumvent them by building your own machine code display routine. (That's something for you to figure out on your own.)

PROGRAM 13.1A Linearity Pattern Generator

```
10 FOR I = 0 TO 43 STEP 7
20 FOR N = 0 TO 63 STEP 7
30 PLOT N,I
40 NEXT N
50 NEXT I
60 PAUSE 32000
```

If you have a color TV/monitor, then you can adjust the color purity with the one-line program 13.1B. It creates an all-white screen for over 10 minutes and then terminates itself. For longer display periods, put the **PAUSE** in a **FOR-NEXT** loop. If any tinting is present when you run it, then the color purity needs adjusting. The program will halt early when you press any key.

PROGRAM 13.1B Color Purity (Blank Screen) Generator

```
10 PAUSE 32000
```

Last, you can adjust convergence on a color TV/monitor using 13.1C. The dots this program generates are only 2×2 points in size and are white against a black field. You could also use this program for linearity, centering, height, and width adjustments, and you might just prefer it. The border makes centering easier, but the more widely spaced dots of the first program appear easier to use for linearity checks.

PROGRAM 13.1C Convergence Generator

```
10 FOR I = 0 TO 23
20 POKE 16418,0
30 PRINT AT I,0; "(32 graphic periods)"
40 NEXT I
50 PAUSE 32000
```

Perhaps more importantly, Program 13.1C demonstrates a technique for printing on the last two lines of the screen. Location 16418 contains an integer variable, DF_SZ, which determines the number of lines reserved for the lower portion of the screen. Normally that number is two, but it can be changed within a running program if the system does not need to use the lower part of the screen (if the system does need access, a crash may result). We have told the computer not to reserve any lines at the bottom of the screen and so we can print a full 24 lines (count them). DF_SZ can also be increased to reserve more than two lines at the bottom. Since the **SCROLL** command rotates only the upper portion of the screen, increasing DF_SZ can be used to effect a partial screen **SCROLL** entirely in BASIC. Simply **POKE** 16418 with the number of lines at the bottom of the screen you don't want to scroll and then execute a **SCROLL** command.

Now that you've reached the end of this book, you should be knowledgeable about the inner workings of your Timex Sinclair. You may be living in an environment controlled to perfection by your computer or you may have lots of ideas on how you are going to control things around you in the future. Either way, we hope you have had fun with this book and that it has helped you to create some useful devices. Drop us a line and let us know how your projects turned out or tell us about your own inventions for the Timex Sinclair. Good luck!

8-Bit Binary/Decimal Conversion

MOST SIG BITS	LEAST SIGNIFICANT BITS															
	0	0	0	0	0	0	0	0	1	1	1	1	1	1	1	1
	0	0	0	0	1	1	1	1	0	0	0	0	1	1	1	1
	0	0	1	1	0	0	1	1	0	0	1	1	0	0	1	1
	0	1	0	1	0	1	0	1	0	1	0	1	0	1	0	1
0000	0	1	2	3	4	5	6	7	8	9	10	11	12	13	14	15
0001	16	17	18	19	20	21	22	23	24	25	26	27	28	29	30	31
0010	32	33	34	35	36	37	38	39	40	41	42	43	44	45	46	47
0011	48	49	50	51	52	53	54	55	56	57	58	59	60	61	62	63
0100	64	65	66	67	68	69	70	71	72	73	74	75	76	77	78	79
0101	80	81	82	83	84	85	86	87	88	89	90	91	92	93	94	95
0110	96	97	98	99	100	101	102	103	104	105	106	107	108	109	110	111
0111	112	113	114	115	116	117	118	119	120	121	122	123	124	125	126	127
1000	128	129	130	131	132	133	134	135	136	137	138	139	140	141	142	143
1001	144	145	146	147	148	149	150	151	152	153	154	155	156	157	158	159
1010	160	161	162	163	164	165	166	167	168	169	170	171	172	173	174	175
1011	176	177	178	179	180	181	182	183	184	185	186	187	188	189	190	191
1100	192	193	194	195	196	197	198	199	200	201	202	203	204	205	206	207
1101	208	209	210	211	212	213	214	215	216	217	218	219	220	221	222	223
1110	224	225	226	227	228	229	230	231	232	233	234	235	236	237	238	239
1111	240	241	242	243	244	245	246	247	248	249	250	251	252	253	254	255

DECIMAL/BINARY EQUIVALENTS, 0 TO 255

1000	−128	−127	−126	−125	−124	−123	−122	−121	−120	−119	−118	−117	−116	−115	−114	−113
1001	−112	−111	−110	−109	−108	−107	−106	−105	−104	−103	−102	−101	−100	−99	−98	−97
1010	−96	−95	−94	−93	−92	−91	−90	−89	−88	−87	−86	−85	−84	−83	−82	−81
1011	−80	−79	−78	−77	−76	−75	−74	−73	−72	−71	−70	−69	−68	−67	−66	−65
1100	−64	−63	−62	−61	−60	−59	−58	−57	−56	−55	−54	−53	−52	−51	−50	−49
1101	−48	−47	−46	−45	−44	−43	−42	−41	−40	−39	−38	−37	−36	−35	−34	−33
1110	−32	−31	−30	−29	−28	−27	−26	−25	−24	−23	−22	−21	−20	−19	−18	−17
1111	−16	−15	−14	−13	−12	−11	−10	−9	−8	−7	−6	−5	−4	−3	−2	−1

TWO'S COMPLEMENT EQUIVALENTS, −1 TO −128
(0 TO +127 REMAIN THE SAME)

8-Bit Hexadecimal/ Decimal Conversion

MOST SIG DIGIT	LEAST SIGNIFICANT BITS															
	0	1	2	3	4	5	6	7	8	9	A	B	C	D	E	F
0	0	1	2	3	4	5	6	7	8	9	10	11	12	13	14	15
1	16	17	18	19	20	21	22	23	24	25	26	27	28	29	30	31
2	32	33	34	35	36	37	38	39	40	41	42	43	44	45	46	47
3	48	49	50	51	52	53	54	55	56	57	58	59	60	61	62	63
4	64	65	66	67	68	69	70	71	72	73	74	75	76	77	78	79
5	80	81	82	83	84	85	86	87	88	89	90	91	92	93	94	95
6	96	97	98	99	100	101	102	103	104	105	106	107	108	109	110	111
7	112	113	114	115	116	117	118	119	120	121	122	123	124	125	126	127
8	128	129	130	131	132	133	134	135	136	137	138	139	140	141	142	143
9	144	145	146	147	148	149	150	151	152	153	154	155	156	157	158	159
A	160	161	162	163	164	165	166	167	168	169	170	171	172	173	174	175
B	176	177	178	179	180	181	182	183	184	185	186	187	188	189	190	191
C	192	193	194	195	196	197	198	199	200	201	202	203	204	205	206	207
D	208	209	210	211	212	213	214	215	216	217	218	219	220	221	222	223
E	224	225	226	227	228	229	230	231	232	233	234	235	236	237	238	239
F	240	241	242	243	244	245	246	247	248	249	250	251	252	253	254	255

DECIMAL/HEXIDECIMAL EQUIVALENTS, 0 TO 255

	0	1	2	3	4	5	6	7	8	9	A	B	C	D	E	F
8	−128	−127	−126	−125	−124	−123	−122	−121	−120	−119	−118	−117	−116	−115	−114	−113
9	−112	−111	−110	−109	−108	−107	−106	−105	−104	−103	−102	−101	−100	−99	−98	−97
A	−96	−95	−94	−93	−92	−91	−90	−89	−88	−87	−86	−85	−84	−83	−82	−81
B	−80	−79	−78	−77	−76	−75	−74	−73	−72	−71	−70	−69	−68	−67	−66	−65
C	−64	−63	−62	−61	−60	−59	−58	−57	−56	−55	−54	−53	−52	−51	−50	−49
D	−48	−47	−46	−45	−44	−43	−42	−41	−40	−39	−38	−37	−36	−35	−34	−33
E	−32	−31	−30	−29	−28	−27	−26	−25	−24	−23	−22	−21	−20	−19	−18	−17
F	−16	−15	−14	−13	−12	−11	−10	−9	−8	−7	−6	−5	−4	−3	−2	−1

NEGATIVE DECIMAL/HEXADECIMAL EQUIVALENTS, −1 TO −128 (0 TO +127 REMAIN THE SAME)

Appendix C

Powers of Two

$$2**0=1 \qquad 2**10=1024 \qquad (1K)$$
$$2**1=2 \qquad 2**11=2048 \qquad (2K)$$
$$2**2=4 \qquad 2**12=4096 \qquad (4K)$$
$$2**3=8 \qquad 2**13=8192 \qquad (8K)$$
$$2**4=16 \qquad 2**14=16384 \qquad (16K)$$
$$2**5=32 \qquad 2**15=32768 \qquad (32K)$$
$$2**6=64 \qquad 2**16=65536 \qquad (64K)$$
$$2**7=128 \qquad 2**17=131072 \qquad (128K)$$
$$2**8=256 \qquad 2**18=262144 \qquad (256K)$$
$$2**9=512 \qquad 2**19=524288 \qquad (512K)$$

American Standard Code for Information Interchange (ASCII)

Character or command	Decimal code	Hex code	Character or command	Decimal code	Hex code	Character or command	Decimal code	Hex code
NUL	0	0	+	43	2B	X	88	58
SOH	1	1	,	44	2C	Y	89	59
STX	2	2	–	45	2D	Z	90	5A
ETX	3	3	.	46	2E	[	91	5B
EOT	4	4	/	47	2F	\	92	5C
ENQ	5	5	0	48	30	]	93	5D
ACK	6	6	1	49	31		94	5E
BELL	7	7	2	50	32	³	95	5F
BS	8	8	3	51	33		96	60
HT	9	9	4	52	34	a	97	61
LF	10	A	5	53	35	b	98	62
VT	11	B	6	54	36	c	99	63
FF	12	C	7	55	37	d	100	64
CR	13	D	8	56	38	e	101	65
SO	14	E	9	57	39	f	102	66
SI	15	F	:	58	3A	g	103	67
DLE	16	10	;	59	3B	h	104	68
XON (or DC1)	17	11	<	60	3C	i	105	69
			=	61	3D	j	106	6A
DC2	18	12	>	62	3E	k	107	6B
XOFF (or DC3)	19	13	?	63	3F	l	108	6C
			@	64	40	m	109	6D
DC4	20	14	A	65	41	n	110	6E
NAK	21	15	B	66	42	o	111	6F
SYN	22	16	C	67	43	p	112	70
ETB	23	17	D	68	44	q	113	71
CAN	24	18	E	69	45	r	114	72
EM	25	19	F	70	46	s	115	73
SUB	26	1A	G	71	47	t	116	74
ESC	27	1B	H	72	48	u	117	75
FS	28	1C	I	73	49	v	118	76
GS	29	1D	J	74	4A	w	119	77

Character or command	Decimal code	Hex code	Character or command	Decimal code	Hex code	Character or command	Decimal code	Hex code
RS	30	1E	K	75	4B	x	120	78
US	31	1F	L	76	4C	y	121	79
SPACE	32	20	M	77	4D	z	122	7A
!	33	21	N	78	4E	{	123	7B
"	34	22	O	79	4F	|	124	7C
#	35	23	P	80	50	}	125	7D
$	36	24	Q	81	51		126	7E
%	37	25	R	82	52	DEL	127	7F
&	38	27	S	83	53			
	39	27	T	84	54			
(	40	28	U	85	55			
)	41	29	V	86	56			
*	42	2A	W	87	57			

Symbols Used in This Book

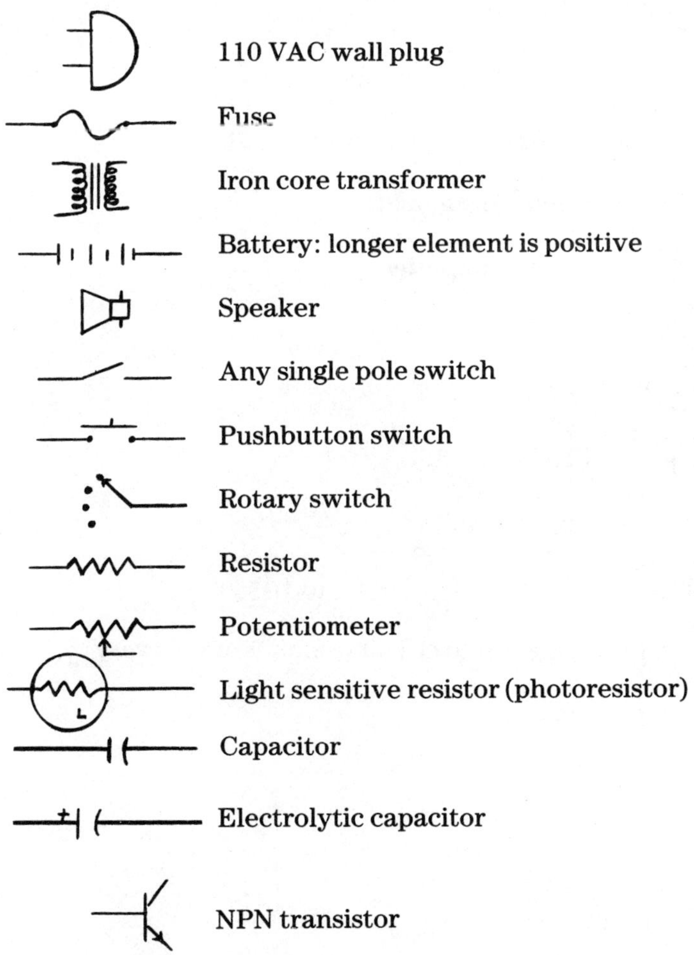

110 VAC wall plug

Fuse

Iron core transformer

Battery: longer element is positive

Speaker

Any single pole switch

Pushbutton switch

Rotary switch

Resistor

Potentiometer

Light sensitive resistor (photoresistor)

Capacitor

Electrolytic capacitor

NPN transistor

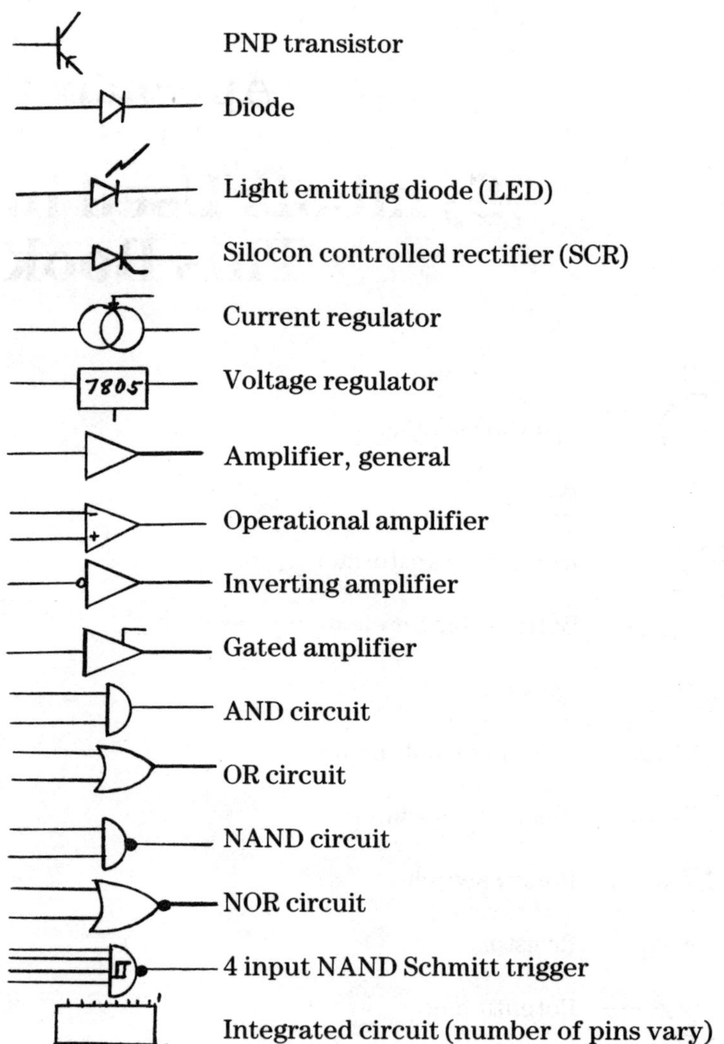

PNP transistor

Diode

Light emitting diode (LED)

Silocon controlled rectifier (SCR)

Current regulator

Voltage regulator

Amplifier, general

Operational amplifier

Inverting amplifier

Gated amplifier

AND circuit

OR circuit

NAND circuit

NOR circuit

4 input NAND Schmitt trigger

Integrated circuit (number of pins vary)

Appendix F

Alphabetic and Numeric Listing of Assembly Language Commands Used in This Book

Alphabetic			Numeric		
Command	Decimal Code	Hex Code	Command	Hex Code	Decimal Code
ADD HL,DE	25	19	NOP	00	0
AND C	161	A1	INC BC	03	3
AND n	230,n	E6,n	LD B,n	06	6
CALL qp	205,p,q	CD,p,q	DEC C	0D	13
CP,n	254,n	FE,n	LD C,n	0E,n	14,n
CPL	47	2F	DJNZ e	10,e−2	16,e−2
DEC C	13	0D	LD DE,mn	11,n,m	17,n,m
DJNZ e	16,e−2	10,e−2	RLA	17	23
IN A,254	219,254	DB,FE	JR,e	18,e−2	24,e−2
INC BC	3	03	ADD HL,DE	19	25
INC HL	35	23	LD E,n	1E,n	30,n
JR C,e	56,e−2	38,e−2	RRA	1F	31
JR e	24,e−2	18,e−2	JR NZ,e	20,e−2	32,e−2
JR NC,e	48,e−2	30,e−2	LD HL,mn	21,n,m	33,n,m
JR NZ,e	32,e−2	20,e−2	INC HL	23	35
JR Z,e	40,e−2	28,e−2	JR Z,e	28,e−2	40,e−2
LD A,n	62,n	3E,n	CPL	2F	47
LD B,n	6,n	06,n	JR NC,e	30,e−2	48,e−2
LD C,n	14,n	0E,n	JR C,e	38,e−2	56,e−2
LD E,n	30,n	1E,n	LD A,n	3E,n	62,n
LD B,H	68	44	LD B,H	44	68
LD C,A	79	4F	LD C,L	4D	77
LD C,L	77	4D	LD C,A	4F	79
LD DE,mn	17,n,m	11,n,m	LD (HL),A	77	119
LD HL,mn	33,n,m	21,n,m	AND C	A1	161
LD (HL),A	119	77	RET	C9	201
NOP	0	00	RR C	CB,19	203,25
OUT 255,A	211,255	D3,FF	CALL qp	CD,p,q	205,p,q
RET	201	C9	OUT 255,A	D3,FF	211,255
RET C	216	D8	RET C	D8	216
RLA	23	17	IN A,254	DB,FE	219,254
RRA	31	1F	AND n	E6,n	230,n
RR C	203,25	CB,19	CP n	FE,n	254,n

International Morse Code

A	.—	Z	——..
B	—...	1	.————
C	—.—.	2	..———
D	—..	3	...——
E	.	4	....—
F	..—.	5	.....
G	——.	6	—....
H	....	7	——...
I	..	8	———..
J	.———	9	————.
K	—.—	0	—————
L	.—..	PERIOD	.—.—.—
M	——	COMMA	——..——
N	—.	?	..——..
O	———	DOUBLE DASH	—...—
P	.——.	COLON	———...
Q	——.—	SEMICOLON	—.—.—.
R	.—.	(or)	—.——.—
S	...	/	—..—.
T	—	ERROR	....
U	..—	WAIT	.—...
V	...—	EOM	.—.—.
W	.——	EOW	...—.—
X	—..—	INVITATION TO	
Y	—.——	TRANSMIT	—.—

Dot is one unit long; dash is three units; pause between dots and dashes
is one unit; pause between characters is three units; pause between
words is seven units.

Samples Tunes for Program 7.7

Change the 50 in Program 7.7 line 20 to 52 and

LET A$ = ''CJAJ8JAJCJCJCKAJAJAKCJCJCKCJAJ8JA
JCJCJCJCJAJAJAJCJAJ8L''

MARY HAD A LITTLE LAMB

Change the 50 in Program 7.7 line 20 to 62 and

LET A$ = ''3J7J8JAM3J7J8JAM3J7J8JAK7K3K7K5M7J7J3L3
J7KAKAJ8M7J8JAK7K3K5K3M''

WHEN THE SAINTS GO MARCHING IN

Appendix I

Screen Images of Programs in Book

The following pages contain copies of the programs listed throughout the body of this book. The difference between these and the previous program listings is that these listings are exact screen images of the programs as they will appear on your TV or monitor as you type them into the computer. They were produced using the TS 2040 printer and the LLIST command.

```
8000 FOR I=16514 TO 32000
8010 INPUT A$
8020 POKE I,VAL A$
8030 PRINT I,PEEK I
8040 IF I>16534 THEN SCROLL
8050 NEXT I
```

FIGURE 3.1A Machine Code Loader

```
9000 FOR I=16514 TO 32000
9010 IF PEEK I=118 THEN STOP
9020 PRINT I,PEEK I
9030 IF INKEY$<>"" THEN PAUSE 32
000
9040 IF I>16534 THEN SCROLL
9050 NEXT I
```

FIGURE 3.1B Machine Code Reader

```
  10 REM ) COPY COPY 5XPI;5 CLEA
R TAN
  20 PRINT AT 11,10;B$(H)+":"+B$
(M)+":"+B$(S)
  30 IF H=X AND M=Y AND S=Z THEN
 PRINT AT 13,9;"TIME IS UP"
  40 IF INKEY$<>"" THEN PRINT AT
 13,9;"            "
  50 LET S=S+1
  60 IF S=61 THEN GOSUB 110
  90 RAND USR 16514
 100 GOTO 20
 110 LET S=1
 120 LET M=M+1
 130 IF M=61 THEN GOSUB 150
 140 RETURN
 150 LET M=1
 160 LET H=H+1
 170 IF H=25 THEN LET H=1
 180 RETURN
 300 LET X=0
 310 PRINT AT 9,7;"WANT ALARM?Y/
N"
 320 INPUT A$
 330 IF A$<>"Y" THEN GOTO 440
 340 PRINT AT 9,7;"SET ALARM
 "
 350 PRINT AT 11,7;"ENTER HOUR"
 360 INPUT X
 370 PRINT AT 11,13;"MINUTE"
 380 INPUT Y
 390 PRINT AT 11,13;"SECOND"
 400 INPUT Z
 410 LET X=X+1
 420 LET Y=Y+1
 430 LET Z=Z+1
 440 PRINT AT 9,7;"SET TIME
 "
 450 PRINT AT 11,7;"ENTER HOUR
 "
 460 INPUT H
 470 PRINT AT 11,13;"MINUTE"
 480 INPUT M
 490 LET H=H+1
 500 LET M=M+1
 510 LET S=1
 520 CLS
 530 PRINT AT 11,10;"START ME"
 540 IF INKEY$<>"" THEN GOTO 20
 550 GOTO 540
1000 FAST
1010 DIM B$(60,2)
```

```
1020 FOR I=1 TO 60
1030 LET B$(I)=STR$ (I-1)
1040 IF I<11 THEN LET B$(I)="0"+
STR$ (I-1)
1050 NEXT I
1060 SLOW
1070 GOTO 300
```

FIGURE 6.1 24-Hour Clock/Timer

```
 10 REM ▪:( RETURN TAN 6789
 20 FOR A=28 TO 37
 30 POKE 16894,A
 40 FOR B=28 TO 37
 50 POKE 16895,B
 60 FOR C=28 TO 33
 70 FOR D=28 TO 37
 80 FOR E=28 TO 37
 90 POKE 16897,C
100 POKE 16898,D
110 POKE 16900,E
120 RAND USR 16514
130 NEXT E
140 NEXT D
150 NEXT C
160 NEXT B
170 NEXT A
180 STOP
200 PRINT AT 11,10;"ST:AR.T"
210 IF INKEY$<>"" THEN GOTO 20
220 GOTO 210
```

FIGURE 6.2 Tenth-Second Stopwatch

```
10 REM 2 COPY LN 3▪/ RAND TAN
20 RAND USR 16514
```

FIGURE 7.1 SAVE-Like Signal

```
10 REM Y▪<= RETURN 350R  ▪?( R
ETURN PEEK COPY ▪?( RETURN / GOS
UB
20 FAST
30 RAND USR 16514
```

PROGRMA 7.2 Continuous Tone Generator

```
   10 REM Y <= RETURN 35OR NEW ?
RETURN ?O POKE  ?( RETURN PEEK C
OPY    ?( RETURN / FAST
   20 FAST
   30 RAND USR 16514
   40 SLOW
```

FIGURE 7.3 Morse Code Oscillator

```
   10 REM ) COPY COPY 5 PEEK COP
Y  ?( RETURN <= RETURN  ?( RETUR
N  ;5 LIST TAN
   20 PRINT "INPUT FREQ NUMBER"
   30 INPUT N
   40 PRINT "INPUT LENGTH NUMBER"
   50 INPUT HILO
   60 POKE 16523,N+2
   70 POKE 16529,N
   80 POKE 16519,INT (HILO/256)
   90 POKE 16518,HILO-256*INT (HI
LO/256)
  100 IF INKEY$<>"" THEN RAND USR
16514
  110 GOTO 100
```

FIGURE 7.4 Tone Burst Generator

```
   10 REM ) COPY COPY 5 PEEK COP
Y  ?( RETURN <= RETURN  ?( RETUR
N  ;5 LIST  =5 EY <= RETURN 35OR
 ;5 RUN ( PAUSE /<=
   20 PRINT AT 11,7;"INPUT SEC/BE
AT"
   30 INPUT R
   40 IF R<.1 OR R>12.8 THEN GOTO
 90
   50 POKE 16537,INT (.5+R/.05)
   60 FAST
   70 RAND USR 16514
   80 STOP
   90 PRINT "VALUE OUT OF RANGE.T
RY AGAIN"
  100 GOTO 20
```

FIGURE 7.5 Metronome

```
   10 REM ) COPY COPY 5 ■PEEK COP
Y ■0( RETURN <= RETURN ■M( RETUR
N  ;S LIST TAN
   20 IF INKEY$="" THEN GOTO 20
   30 LET Y=INT (-2.5+142.04545*1
.0596431**-(CODE INKEY$-28))
   40 POKE 16523,Y+2
   50 POKE 16529,Y
   60 RAND USR 16514
   70 GOTO 20
```

FIGURE 7.6 Keyboard Music

```
   10 REM ) COPY COPY 5 ■PEEK COP
Y ■7( RETURN <= RETURN ■7( RETUR
N  ;S LIST TAN
   20 FOR I=1 TO 50 STEP 2
   30 LET Y=INT (-2.5+142.04545*1
.0596431**-(CODE A$(I)-28))
   40 POKE 16523,Y+2
   50 POKE 16529,Y
   60 POKE 16519,CODE A$(I+1)-46
   70 RAND USR 16514
   80 NEXT I
```

FIGURE 7.7 Music from Memory (Player Piano)

```
   10 REM Y LOAD <= RETURN ?■ TAN

   20 LET X=USR 16514
   30 PRINT X;" ";
   40 GOTO 20
```

FIGURE 8.1 Five-Key Reader

```
   10 REM 5■RND■ Y LOAD <= RETURN
J?Y■■77AC$ ;( RUN TAN
   20 LET X=63
   30 LET Y=41
   40 PLOT X,Y
   50 RAND USR 16514
   60 GOTO 71-PEEK 16535
   70 UNPLOT X,Y
```

```
   71 LET X=X+PEEK 16537-PEEK 165
36
   80 LET Y=Y+PEEK 16538-PEEK 165
39
   90 IF INKEY$="C" THEN GOSUB 13
0
  100 IF INKEY$="S" THEN SAVE "DR
AW"
  110 GOTO 40
  130 INPUT R
  140 FOR J=0 TO 2*R STEP .2
  150 LET Z=INT (.5+SQR (2*R*J-J*
*2))
  160 LET Q=INT (.5+X-R+J)
  170 PLOT Q,Y+Z
  180 PLOT Q,Y-Z
  190 NEXT J
  200 RETURN
```

FIGURE 8.2 Draw

```
  10 REM <= RETURN ?* TAN
  20 LET X=USR 16514
  30 LET X=X-33
  40 LET K4=INT (X/16)
  50 LET K3=INT (-2*K4+X/8)
  60 LET K2=INT (-4*K4-2*K3+X/4)
  70 LET K1=INT (-8*K4-4*K3-2*K2
+X/2)
  80 PRINT AT 9,0;K4,K3,K2,K1
  90 GOTO 20
```

FIGURE 10.1 Four-Input Reader and Decoder

```
  10 REM )* 5  ??<= RETURN ;***K
 RAND <= RETURN ;***S RAND / LO
AD
  20 FAST
  30 LET X=USR 16514
  40 LET F=INT (.5+(X/(65535*43+
8*X))*3.23*1000000)
  50 PRINT AT 11,10;F/1000;" KHZ
"
  60 SLOW
  70 FOR I=1 TO 50
  80 NEXT I
  90 CLS
 100 GOTO 20
```

FIGURE 10.2 Frequency Counter

```
  10 REM ) COPY COPY Y█<= RETURN
 380R 57 ;S CLEAR PEEK COPY 57 ;
S CLEAR / REM
  20 PRINT "INPUT N1"
  30 INPUT N1
  40 PRINT "INPUT N2"
  50 INPUT N2
  60 POKE 16525,INT (N1/256)
  70 POKE 16524,N1-256*INT (N1/2
56)
  80 POKE 16533,INT (N2/256)
  90 POKE 16532,N2-256*INT (N2/2
56)
 100 FAST
 110 RAND USR 16514
```

FIGURE 11.1 Pulse Generator

```
  10 FOR I=0 TO 42 STEP 7
  20 FOR N=0 TO 63 STEP 7
  30 PLOT N,I
  40 NEXT N
  50 NEXT I
  60 PAUSE 32000
```

FIGURE 13.1A Linearity Pattern Generator

```
  10 PAUSE 32000
```

FIGURE 13.1B Color Purity (Blank Screen) Generator

```
  10 FOR I=0 TO 23
  20 POKE 16418,0
  30 PRINT AT I,0;"████████████████
████████████████"
  40 NEXT I
  50 PAUSE 32000
```

FIGURE 13.1C Convergence Generator

Index